A City in Blue and Green

PETER G. ROWE

Raymond Garbe Professor of Architecture and Urban Design
Harvard University Distinguished Service Professor
Graduate School of Design, Harvard University

LIMIN HEE

Director of Research
Centre for Liveable Cities (CLC)
Singapore

ISBN 978-981-13-9596-3
ISBN 978-981-13-9597-0 (eBook)
https://doi.org/10.1007/978-981-13-9597-0

This Springer imprint is published by the registered company Springer Nature Singapore Pte Ltd.
The registered company address is: 152 Beach Road, #21-01/04 Gateway East, Singapore 189721, Singapore

To the old and new pioneers of Singapore.

TABLE OF CONTENTS

PREFACE

Peter Rowe and I have known each other for a long time. We collaborated through design studios between the Singapore Urban Redevelopment Authority and the Harvard Graduate School of Design as early as the 1990s. Limin, now director of research at my centre, was also his doctoral student at Harvard. Peter often fondly recalls his childhood in Hong Kong, and his frequent visits to Singapore. It is of no surprise that he continues to be a keen observer of developments in Asia, having many former students in the region, and especially in China, where he is also Visiting Professor at Tsinghua University.

When we invited Peter to our Centre as a Visiting Fellow back in 2015, he noted with admiration Singapore's transformation over the years, especially in how we have developed but retained a green mantle as a distinctive part and parcel of our cityscape. I also shared with him some of our programmes to beautify our waterways, notably, the Active, Beautiful and Clean Waters Programme. In our conversation, Peter saw an opportunity to capture the subject of Singapore's integral use of blue and green infrastructure as part of our recent history of planning in a book. He expressed a keenness to write about our success in greening the city and water management as he felt there are applicable lessons there for other cities. Such a book would also help to present the model of Singapore's planning, which has great potentials beyond that of a city in a garden, to an international audience of practitioners and researchers.

Peter then very kindly took up the invitation to write this book in collaboration with our researchers at the Centre, and in the process, engaging with many of Singapore's urban pioneers, practitioners in both our public and private sectors, as well as visiting several projects to gain insights into how we have created a city in blue and green.

I would like to thank Peter for putting this book together, having devoted much of his precious time and energy on this collaboration. And to thank the many who have shared their experiences and thoughts with Peter.

Khoo Teng Chye
Centre for Liveable Cities, 2019

LIST OF ILLUSTRATIONS

chapter

01 INTRODUCTION

P. G. Rowe and L. Hee, *A City in Blue and Green*,
https://doi.org/10.1007/978-981-13-9597-0_1

This book is about Singapore's development into a city in which water and vegetation, along with associated environmental, technical, social and political aspects have been harnessed and cultivated into a livable sustainable way of life. It is also a story about a unique and thoroughgoing approach to large-scale and potentially transferable water sustainability, within largely urbanized circumstances, which can be achieved, along with complementary roles of environmental conservation, ecology, public open-space management and the greening of buildings, together with infrastructural improvements.

1. SATELLITE IMAGE OF SINGAPORE

a.

ASPECTS OF 'BLUE' AND 'GREEN'

In the context of this book, 'blue and green' or 'blue-green' network planning and programs seek to protect the hydrologic and ecological values of Singapore's urban and other landscapes and to provide resilient measures to address threats and environmental degradation, such as water dependency on others and deforestation. Overall, 'blue-green' planning has become increasingly observed as governments and the entities involved are subject to unprecedented rates and volumes of urban growth, coupled with loss of habitat and biodiversity. According to international expert panels the next two decades are likely to present wider opportunities for mitigation efforts and the instigation of sustainable ecosystem practices. Even sizeable international institutions such as UN-Habitat are often involved in supporting such practices, particularly in collaboration with local municipalities and government organizations. Generally, 'blue-green' networks expand the rehabilitation of the 'blue' water cycle within urban areas, as well as complementing conventional engineering solutions, or so-called 'grey networks.[1] Singapore's recent Active, Beautiful, Clean Waters Programme (ABC Waters Programme) is a strong example of this approach.

More specifically,' blue-green' networks consist of 'blue' water-based elements,' green' vegetated-based elements, 'green' technologies and often low-carbon and climate resistant infrastructure. There, 'blue' elements usually comprise rivers, streams, storm-water drains, irrigation channels, canals, wetlands, freshwater bodies and swales. 'Green' components usually include: roadside trees, recreation zones, playgrounds, parks, forests, greenways and riparian strips. Larger spatial organizations range from catchments to sub-catchments at a neighborhood level, and to micro-catchments at urban block levels. By way of orientation, 'blue-green' networks seek to recreate a natural water cycle while contributing to the amenity of urban areas through water management and green infrastructure together.[2] In short, the aim is to contribute to and protect hydrologic and ecological values of urban landscapes, while providing resilient and adaptive measures to address future changes in environmental conditions and related activities.

2. THE GREEN AND BLUE PLAN ARISING FROM THE 1991 CONCEPT PLAN

b. SINGAPOREAN SIGNIFICANCE OF 'BLUE AND GREEN'

When one thinks of Singapore and what it produces economically as a nation state, one cannot point to a particular line of automobiles, consumer goods or other products. Its Gross Domestic Product (GDP) is high and heavily weighted towards the service sector with around 73.4 percent and industry with the remaining 26.6 percent of the total.[3] It is, after all, a highly-developed, free-market economy that is very open to the world and with little corruption, low tax rates, as well as being the most pro-business and third highest GDP per capita in purchasing parity terms in the world. Its government-linked companies, like Singapore Airlines, PSA Corporation Limited, Sing Tel, and the ST Engineering and Media Corporation play significant roles, as do its property development enterprises like Capitaland and Keppel Land. Major sources of revenue are electronics, chemicals, and, of course, services. Singapore also relies on an extended concept of intermediary trade in these regards, by purchasing raw goods and refining them for export, through one of the busiest harbors in the world. In this entrepôt-like tradition gross exports amount to about $330 billion (2016) in pharmaceuticals, petroleum products and equipment, with inputs of about $283 billion (2016), also in equipment, fuels and machinery, alongside of food and consumer goods.[4] A good number of Singaporeans work in the civil service, with a ratio of 1:71.4 of civil servants to population, compared to the U.K. with 1:118, China with 1:108, but with Malaysia with 1:19 and Russia with 1:84.[5] In short Singapore is neither a conspicuous maker of particular products, nor a place with numerous employees on state payrolls. Nor is it conspicuously innovative, largely copying from others or applying rather than inventing leading-edge technologies or manners of doing things. By contrast, it is highly pragmatic in these regards.

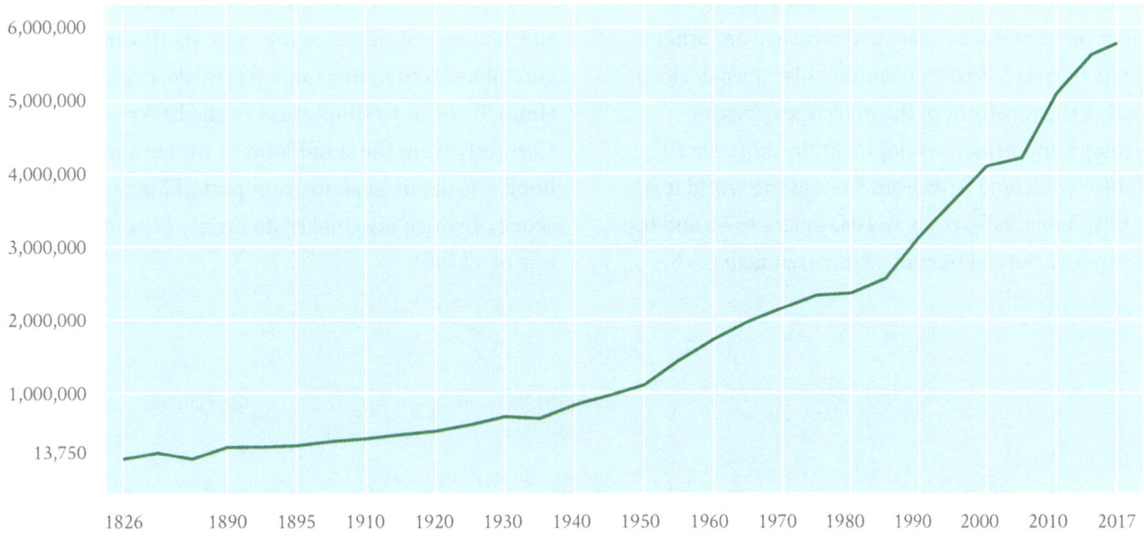

3A. THE POPULATION OF SINGAPORE, 1826–2017*

*Differential recent rates of growth due to immigration of non-resident population.

	TOTAL POPULATION	NON-RESIDENT POPULATION	NON-RESIDENT %
2000	4,027,887	745,524	18.7
2010	5,076,732	1,305,011	25.7
2015	5,540,000	1,600,000	28.9
2030	6,700,000	3,015,000	45.0

3B. SINGAPORE'S TOTAL AND NON-RESIDENT POPULATION

What Singapore is noteworthy for, however, is, in a way, itself and the character and distinction of its city making. It has an unparalleled and extraordinary public housing program supporting over 80 percent of its resident population. It boasts a plethora of entertainment, sports and other venues. It is one of the world's downtowns with respect to shopping and life-style activities. In fact Singapore receives over fifteen million visitors per year, a number that has risen steadily, particularly over the past dozen years and mostly from other parts of Asia.[6] It is environmentally squeaky clean and at the forefront of the marriage between ecology and urban development. In short for all those who come from elsewhere in the world it is a hospitable, safe, clean and easy place to be and to enjoy a variety of pursuits. Underlying all of this attractiveness, however, is an island city-state that is 'green and clean' as advertised as well as resplendent in its aqueous environments. Its liveability and attractiveness, in short, derives strongly from its 'blue-green' environment co-mingling with its urban landscapes. In part this is strategic in the sense of water resource sustainability. However, it is also symbiotic in the sense that 'blue' and 'green' must go together to become such an integral part of Singapore. Moreover, it is for this reason that the 'blue-green' aspect is the most significant contribution that Singapore has made, even if, at times, it seems to simply lurk in the background. Certainly from the standpoint of the authors of this book it is the most significant part of Singapore, the scope of which also makes it entirely distinctive and one of a kind.

	GDP (BN)	GDP/CAP (K)
NEW YORK	1,558	120
TORONTO	305	116
BOSTON	382	81
SEOUL	688	69
LONDON	542	63
SINGAPORE	306	55
HONG KONG	291	40

4. ECONOMIC INTENSITY OF SINGAPORE AMONG SELECTED CITIES IN 2013

c.

ORGANIZATION OF THE BOOK

The remainder of the book is organized into five chapters. Chapter two which follows, deals with the early days of Singapore, dating from early sightings by Chinese mariners and others, to power struggles among competing regimes and the development of Temasek as a relatively early rudimentary settlement on the island. It also spans to the British colonial period beginning with Stamford Raffles and ending with devolution of the British colonial possessions in and around Malaysia after the occupation by the Japanese. This is followed by the break with the Malay Federation and the formation of an independent island city state in 1965. Throughout, while there is a certain amount of historical politically-driven narrative, concentration is also placed on the physical character of Singapore's changing landscape.

Chapter three follows on with Prime Minister Lee Kuan Yew's admonition to Singaporeans to make their environment 'clean and green', as both a practical way of improving the chances of attracting outside foreign investment, achieving a certain equity in environmental quality among a very mixed population and as a metaphor for what Singapore should be like as a place.[7] In this last regard, parallels are drawn with other metaphors shaping urban life elsewhere, such as American pastoralism, the machine in the garden and even Italian versions of city and countryside. Efforts in Singapore to vivify its metaphor around 'blue and green' are also discussed, along with how they have become socialized and made more habitual in Singaporean life.

Chapter four examines the difficult issue of achieving water sustainability for Singapore, through water importation and finally three other 'National Taps'. These include: improvement of water catchment facilities; production of NEWater from effluent and a closed loop system of reticulation and storage; alongside of desalination. Foregrounded is also the radical revision of water capture, treatment, and re-use that has and is occurring in Singapore. While specific technologies may not be uniquely Singaporean, the scope, relative scale and sophistication of implementation is rivalled by few, if any, other applications in the world.

Chapter five, dealing with the 'green' aspect of the 'blue-green' arrangement presents the alibi for water capture and treatment operations in the form of deliberate movement from the idea of a 'garden city' to a 'city in nature' and of a tropical kind. It also covers the harnessing of higher levels of biodiversity to achieve such a 'natural' condition and to facilitate water treatment by way of cleansing sequestration. Increasingly sophisticated management of plant life and public open space in Singapore, including programs in direct connection with active waterways are also described and discussed, along with far-reaching research programs.

Finally, chapter six offers opinion about the relative success of Singapore's 'blue-green' plans and likely follow-on engagements. This will include some identification of what might be seen as existential threats to this success from outside of Singapore. Discussion will also speculate about collective habits of mind and behavior that have been instrumental in leading to success, as well as visionary modes of leadership, perseverance and pragmatism. On the whole there has not been anything lucky or flukey about what has been achieved, leading to the topic of transferability to other circumstances elsewhere in the world and how that might be made successfully.[8]

Although most of the focus of the book is upon 'blue and green' programs in Singapore in the relatively recent times of the past 50 or so years, discussion at times will extend back into the British colonial period, if not before, as suggested by chapter two. This is done to underline several aspects or trends in Singapore's history and development. The first concerns the island state's capacity to re-invent itself, sometimes dramatically and not always to the longer-term good. A second concerns the ending of one era forming the initial conditions for a subsequent round of developments. Part of Singapore's persistence with ideas about the perfectibility of the city, for instance, hinges on a pragmatic reflexivity to prior conditions and events. Finally, it is within the long durée of development that the unique qualities and values of Singapore have been shaped.

endnotes

1. As described by texts like Hoyer,J.;W. Dickhaut; I. Kronawitter and B. Weber. 2011.*Water Sensitive Urban Design* (Hamburg: Jovis). And "Blue-Green Cities" https://en.wikipedia.org/wiki/Blue_Green_Cities.

2. *Op cit.*

3. Singapore Department of Statistics (DOS) on Gross Domestic Product by sector.

4. Singapore Exports 1964-2017. https://tradingeconomics.com/singapore/exports

5. See the Public Service Commission of Singapore and statistics by country in https://en.wikipedia.org/wiki/Category: "Civil_Service_by_Country".

6. Singapore 2014 Tourism Sector Performance, International Visitor Annual Statistics.

7. Lee Kuan Yew, Opening Speech of the "Keep Singapore Clean Campaign", 1 October, 1968.

8. Interview with Dr. Liu Thai Ker, Former CEO of HDB and URA and Chairman of CLC on 18 August, 2017.

chapter

02

EARLY DAYS

P. G. Rowe and L. Hee, *A City in Blue and Green*,
https://doi.org/10.1007/978-981-13-9597-0_2

The story begins against the background of Temasek and pre-colonial occupation to the residue of British colonial development and the birth of the Republic of Singapore in 1965 as a fledgling island nation state in South-east Asia. With early colonial settlement dating from 1819 with the arrival of Stamford Raffles and the East India Company, the primeval forested and vegetated circumstances of Singapore were transformed substantially to make way for plantation and other forms of agricultural production. Well served by its location as a choke point between the South China Sea and the Indian Ocean, the colony thrived largely on entrepôt trade with the Malay hinterland through concentration of lightering, warehousing and convenient association of locals and foreigners, primarily around the Singapore River. Loosely conforming to a plan, originally drawn up by Lieutenant Jackson in 1822, a 7 to 12 kilometer necklace-like arrangement of settlements was stretched along the southern coast and played host to Malays, Chinese, Westerners, Indians, Burgis and others. Gradually moving into the interior of the island, much of the urban development was ad hoc and often very dense, with rudimentary public infrastructure and other public improvements evolving slowly over time. As in other parts of Malaya under British domination, responses to the dictates of agricultural production, climatic and topographic circumstances involved clear cutting, re-planting, crude channelization, and until later, little treatment of the residuals of production or occupation. As time and continued development wore on, not to mention the economic boon of tin and rubber transshipment from around 1900, Singapore became increasingly dependent upon other parts of Malaya for infrastructural support and, particularly, fresh water supply and with water supply agreements dating from 1927 and extending well up into the early 1960s. The economically extractive orientation of colonial rule, though continuing to emphasize Singapore's relatively advantageous trading position, did little for large segments of both indigenous and immigrant populations. Nearing the time of the birth of the Republic, overcrowded and poverty-stricken settlements were common, along with frequent local flooding and inadequate utilities. Environmental quality, certainly by modern standards, was almost universally poor leaving most of the island's inhabitants festering in dirty, and dismal conditions. Indeed, Singapore's first Master Plan was approved in 1958. However the conservative planning provisions in the Master Plan were insufficient for Singapore's development needs. Seeking technical assistance from the United Nations, Singapore built up a more comprehensive urban planning system over the years. In tandem, Singapore also created institutions such as the HDB and URA, and put in place pieces of legislation such as the Land Acquisition Act of 1966, both of which proved to be sound and effective. By 1971, Singapore commissioned the first Concept Plan, a long term strategic plan which was influenced by the "Ring City Concept" proposed earlier by the UN experts, and which would guide Singapore's urban development over the next decades.

5. OUTCROPS AT KEPPEL HARBOR

a.

GEOGRAPHIC SETTING AND CONTROL

One of the earliest accounts of what is now Singapore dates back to one Wang Ta-yuan's first-hand description, when he traveled in Nanyang, or the South China Sea, in the first half of the 14th century. According to Wang, Temasek – Singapore's earliest appellation – was inhabited by some Chinese but also by pirates.[1] He also referred to a trading settlement near Long Ya-men – Dragons Teeth Gate – probably making reference to outcrops at the entrance to Keppel Harbor. A navigator during the early Ming Dynasty, Wang also observed how westbound ships crossing the Sunda Straits in front of Temasek were left untouched, while those moving eastward into the South China Sea were often attacked by numerous and armed boats. Earlier written accounts still came by way of P'u Luo Chung and his rendering of the Malay Pulau Ujong, or island at the end of the Malay Peninsula as far back as the third century A.D.[2] Inhabitants were reported to be primitive cannibals also preying on junks plying their way back and on through the South China Sea. Temasek was a choke point in the former Sunda Shelf that once linked the Malay Peninsula to the islands of modern Indonesia.[3] This location was such that refuge could be sought between the seasonal north-east and south-west monsoonal winds, also facilitating navigation both in and out of the South China Sea from the Indian Ocean to the south and west. Indeed, Temasek and now Singapore enjoyed the advantage of command of sea lanes valuable to trading, good anchorage, and sufficient local resources in the presence of potable water, along with defensible positions. These attributes persisted well into the modern era, becoming a leitmotif of Singapore's comparative advantage.

Later on Temasek found itself in the throes of the power struggle between the Srivijayan Empire to the west and the Javanees Empire of the Majapahit to the east. At one time the Srivijayans occupied significant areas of the Malay Peninsula and nearby islands, with Palembang at its heart on today's Sumatra. Apparently, during an inspection tour of the islands Prince San Nila Utama came upon the shores of Temasek near the estuary of the present Singapore River, also observing what seems to have been a lion.[4] According to the Malay Annals he then founded a city at Kuala Temasek which he called Singapora, or the 'Lion City'.[5] Later, in 1365, Temasek was claimed by the Majapahit and made a vassal state. Retaken by the Srivijayan forces in 1390, Iskander, its leader was driven out of Palembang, its capital, and sought refuge in Temasek. As a city, Temasek faced out to the sea and was walled on its northern side by an embankment of compacted earth about five meters wide and three meters in height, stretching from the sea about 1.5 kilometers around the edge of what became Fort Canning Hill.[6] The Singapore River made up most of the southern side of this defensible redoubt. Internally, it was probably comprised of a kampong with timber and thatched buildings, elevated above the ground for ventilation. Inhabitants of Temasek may have engaged in an early form of entrepôt trade, although corroboration by early visitors like Wang Ta-yuan and the Portuguese Torre Pines was not forthcoming. As the historian C.M. Turnbull concludes: "Temasek was a small outpost of Srivijaya, its people harnessed to the sea power of the Empire or, in more troubled times, exploiting their position to prey on other's trade".[7] In the meantime, Temasek was renamed Singapura, or the Lion City.

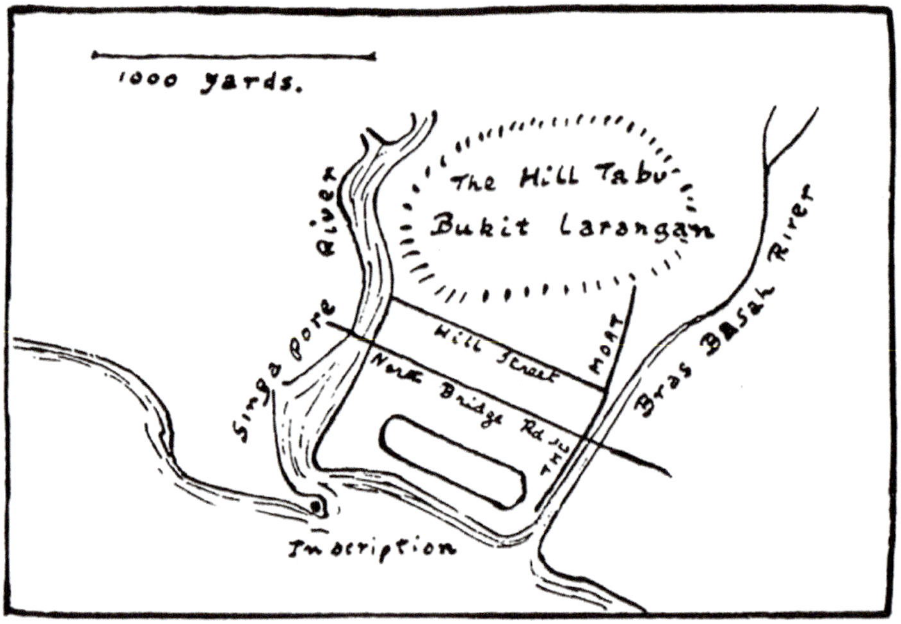

6. PLAN OF TEMASEK

7. HARBOUR AND TOWN OF SINGAPORE

Although not completely deserted after Iskander's flight around 1398, Singapora and the areas around it remained relatively abandoned. It was occupied largely by orang laut or sea nomads who led an existence sometimes trading with passing ships or lying in wait to plunder them.[8] With the fall of Malacca in 1511 to the Portuguese, Singapora became an outpost of the remnant Malaccan Sultanate. This was a position that Singapora held, in spite of the Portugese destruction of the Sultan's naval power in 1507, and successful attacks on Johor. In time the Sultanate retreated in part to Singapora becoming the Temenggong Rulers. For the next 150 to 200 years beyond the sixteenth century, however, the importance of the island dwindled appreciably. This was due to the loss of Malay power in the region leading also to a power vacuum and isolation. The area lost out to Malacca with regard to trade and commerce. In the first decade in the nineteenth century Temenggong's Abdu'r Rahman set up a village on the former site of Singapora, associating himself with a small band of Malays gathered from the orang laut tribes. In total there were perhaps 1000 inhabitants mainly orang laut and small numbers of Chinese, as well as some 20 to 30 Malays in the Temenggong's entourage. At the time this was a typical Malay-ruled village, consisting of the Temenggong's substantial timber house, surrounded by thatched atap houses. Inhabitants grew fruit and depended largely on their livelihood by collecting jungle produce, fishing, small-scale trading and piracy. The most noticeable aspect of the last pursuit was the Golang pirate fleet of 1818 with some 50 ships and about 1,200 men. In addition, there were a number of Chinese engaged in agricultural development in the interior, particularly in 20 to 30 gambier plantations, producing an extract and popular betel nut, with pepper as a supplemental crop. The island was far from pristine in indigenous plantation and wildlife.

8. A MALAYAN STYLE KAMPONG

Less locally, the early nineteenth century world was one of mercantile capitalism with European imperial powers actively engaged in controlling important sea lanes to further their trading interactions especially with East Asia. At first this engaged the Portuguese, trading from Macau through the South China Sea to Malacca and then on to Lisbon via Goa, Mozambique and Angola. Then there were the Dutch in the seventeenth century establishing a headquarters in Batavia on Java and bypassing the Portuguese in Malacca through the Sunda Straits and into the Southern Indian Ocean and before interdicting themselves into Ceylon in 1640 and Malacca in 1641. This enabled them to both dominate the Sunda and Malacca Straits for roughly 150 years. To counter these Dutch ambitions, the British attempted to secure a foothold at the southern end of the Straits of Malacca, proximate to the Sunda Straits and effectively securing sea routes to China and the Malay Archipelago for British trade. In 1819 an expedition led by Stamford Raffles signed a preliminary agreement with Temenggong Abdu'r Rahman the local ruler of Singapora. This was later ratified in the same year in the manner of the doctrine of 'indirect rule' authorizing the British East Indian Company to establish a factory or trading facility in exchange for British protection and 3000 Spanish dollars annually.[9] In effect a lease-hold or 'use right' agreement was consummated. Raffles was an uncommon if controversial administrator. At the time he was the Lieutenant-Governor of Bencoolen, a British possession in Sumatra since 1685, when he abolished slavery among other reforms. He was also an agent of the British East Indian Company with the profit-making orientation that went with the position. More personally, Raffles was an ardent student of local customs and languages and largely possessive of an enlightenment mind-set. However, it wasn't until 1824 with the Anglo-Dutch Treaty in place and with the Residency of John Crawfurd that the British consigned sovereignty to the entire island. Further in 1826, the East India Company united Singapore with Penang, Malacca and several other states to form the Straits Settlements with Singapore as the capital under a Governor. At the time of the first census in 1824 the island's population was counted as 10,683 people.

The 'Town of Singapore' as it was called was represented cartographically by Lieutenant Philip Jackson, at Raffles request, in 1824. Earlier on in 1819 Colonel William Farqhuar, a friend of Raffles and long-serving officer in Malacca and the Riau Archipelago, set about clearing land on the north-east bank of the Singapore River for a market and cantonment of temporary huts.[10] Jackson's plan, overall, was a necklace of communities largely consolidating who and what was in place. Roughly from west to east, there was the Malay Kampong, next to the Chinese settlement on the western side

9. LT. JACKSON'S PLAN OF SINGAPORE IN CONTEXT

of the Singapore River. Across from that was the Government Centre and European Town, trailing off into the Arab and Burgis areas. In area the plan inscribed a coastal zone of 6 or 7 kilometers in length and around 1.5 kilometers in width. Although not entirely according to Raffles wishes, the ethnic separation was how matters were at the time and not untypical of British colonial settlements elsewhere. Major influences in the plan were largely Palladian and Neoclassical, reminiscent of Nash London in 1812.

Organizationally, rectangular grids of streets and roads of various dimensions were deployed along the coast, also with a sizable public space or Padang.[11] Topography also played a role, with Fort Canning Hill dominating the European community, crowned by Government House. British power and wealth was projected through the image of Palladian-style buildings lining the major roads in the plan, especially in comparison to smaller allotments given over in the plan to Chinese shophouses to the southwest. The Singapore River served as the major port of commerce and mercantile activity, involving lightering from and to larger ships at anchor in nearby Keppel Harbor. In short, during the early colonial period, Singapore was a relatively small but growing settlement, increasingly under British hegemony, consisting in form of relatively separated enclaves, although becoming mixed in some cases along the coast. The settlement also had a distinct urban and administrative centre and port facilities, adjoined by looser collections of largely detached dwellings and shophouses.

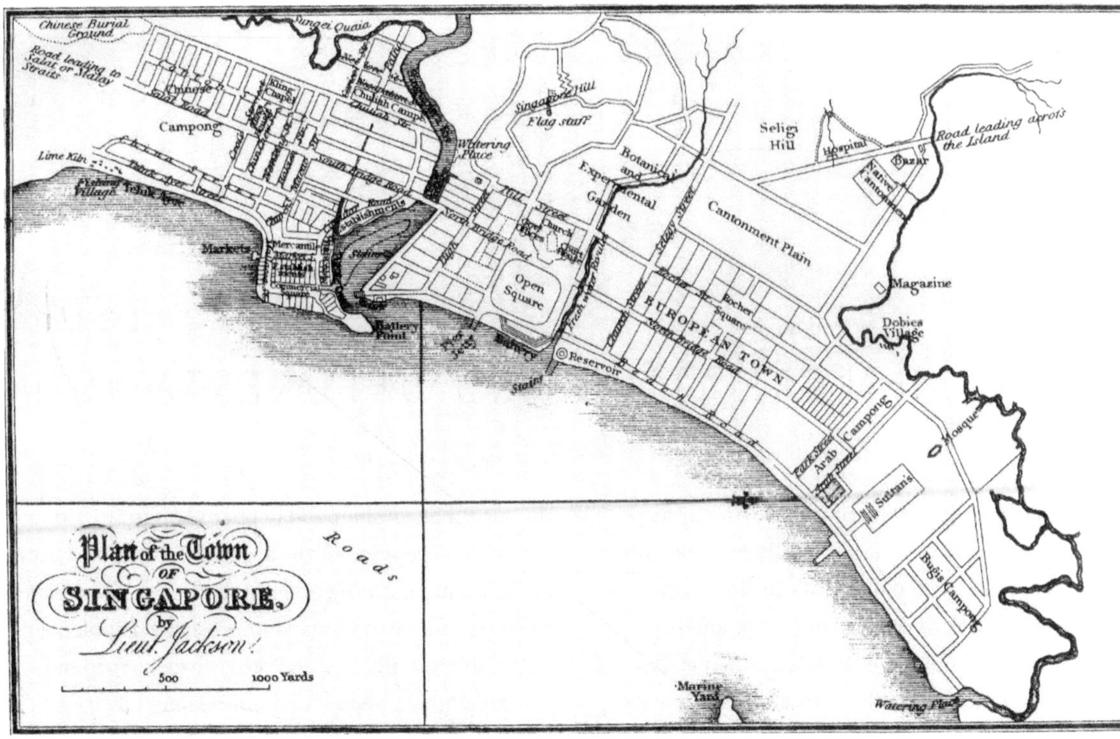

10. A PARTIAL SCHEMATIC OF LT. JACKSON'S PLAN FOR SINGAPORE

11. DEPICTION OF FORT CANNING AND ENVIRONS

b.

TRANSFORMATION AND DEFORESTATION

As mentioned earlier, at the time of Raffles arrival the island of Singapore was not totally pristine. orang laut settlers were located north of the Singapore River, at the mouth of the Kallang River and in the estuaries of the Punggol and Seletar Rivers, as fishing communities mainly, with marginal impacts on the land.[12] At the time the three significant ecological habitats were the primeval forests, the freshwater swamp forests and the mangrove forests. However, by the middle of the nineteenth century the spread of gambier and pepper plantations along with other crops led to the almost total deforestation of the island, with the exception of a few hilltops and steep slopes. Lalang grass and brushcover took over in the absence of crops and other forms of ground cover. Gambier, pepper and nutmeg introduction led to numerous plantations on the interior serving global markets. Then spices were replaced in the earlier twentieth century by rubber plantations up until the Japanese occupation. Also farms to feed

12. A GAMBIER-PEPPER PLANTATION

the population developed even making Singapore self-sufficient in pork, eggs and poultry into the 1980s.[13] Very early on even Raffles decided that potentially valuable land should not be wasted in the construction of a township. After all, spices were the corner stone of most European trade in South East Asia and also behind much of the Anglo-Dutch naval tussle. Also as a part of the British East India Company efforts to gain a foothold in Singapore Raffles established experimental botanic gardens next to his residence at Fort Canning in 1819, leading to many nutmeg plantations until they were wiped out by parasites in the 1850s and 60s.

Early Chaozhou immigrants from Guangdong Province in China also brought agricultural practices and the Kangchu system of social organization to Singapore in the late 1700s. This system revolved around the Kanglar or riverine village headquarters comprised of administrative buildings, general stores, opium dens, and gambling houses, from which radiated out pigsties, vegetable farms and orchards. These Kanglar resulted in a number of bangsals of cleared land for gambier and pepper production.[14] In total, nine to ten men worked a bangsal which became depleted after some fifteen to twenty years, leading to outward expansion of the settlement to create new bangsals. The abandoned areas either reverted to secondary forest or, more likely, to useless lalang grass areas. Rivers and streams were the main avenues for movement into the interior, as roads to Bukit Timah, Kranji and Sarangoon were only established several decades after 1819. In essence, reliance on river transportation meant that kanglars were close to navigable water ways. In short, the development of remote areas of Singapore began along rivers and expanded outwards from central villages. The interstitial locations within this kind of network also provided the broad palm leaves used for roofing, among other purposes.

Both gambier and pepper cultivation and production were destructive in other ways, apart from consuming land area and depleting soils where they stood. Both crops required large amounts of timber for processing through the boiling of leaves and kiln drying requiring extensive amounts of fire wood. This, in turn, led to logging and deforestation. In fact, a report in 1883 showed that a typical bangsal, of the kind described earlier, consumed 2,500 pounds of timber per day, or roughly the same equivalent area of forest as was consumed by the planting of crops in the first place[15]. Certainly by the end of the 1940s and mid-century, the destructive face of gambier and pepper planting were beginning to become clear to Singapore's residents. Another side-effect was the adverse impact on wildlife and birdlife, including virtual extinction of native and rare species like the Oriental Pied Hornbills. Malaysian tigers among larger animals adapted to radically changed vegetated circumstances by beginning to prey on domestic livestock, dogs and even humans. The last tiger was shot apparently in 1930 having been considered legitimate threats from around the 1830s onwards. Today there are no Malaysian tigers in Singapore due to hunting and diminution of habitat. Eventual decline of gambier plantations was due to the lack of timber resources for processing and land title regulations put into effect in the late 1850s to regulate production and falling prices of crops. The upshot also, however, was that the land was exhausted and almost completely deforested by 1900.

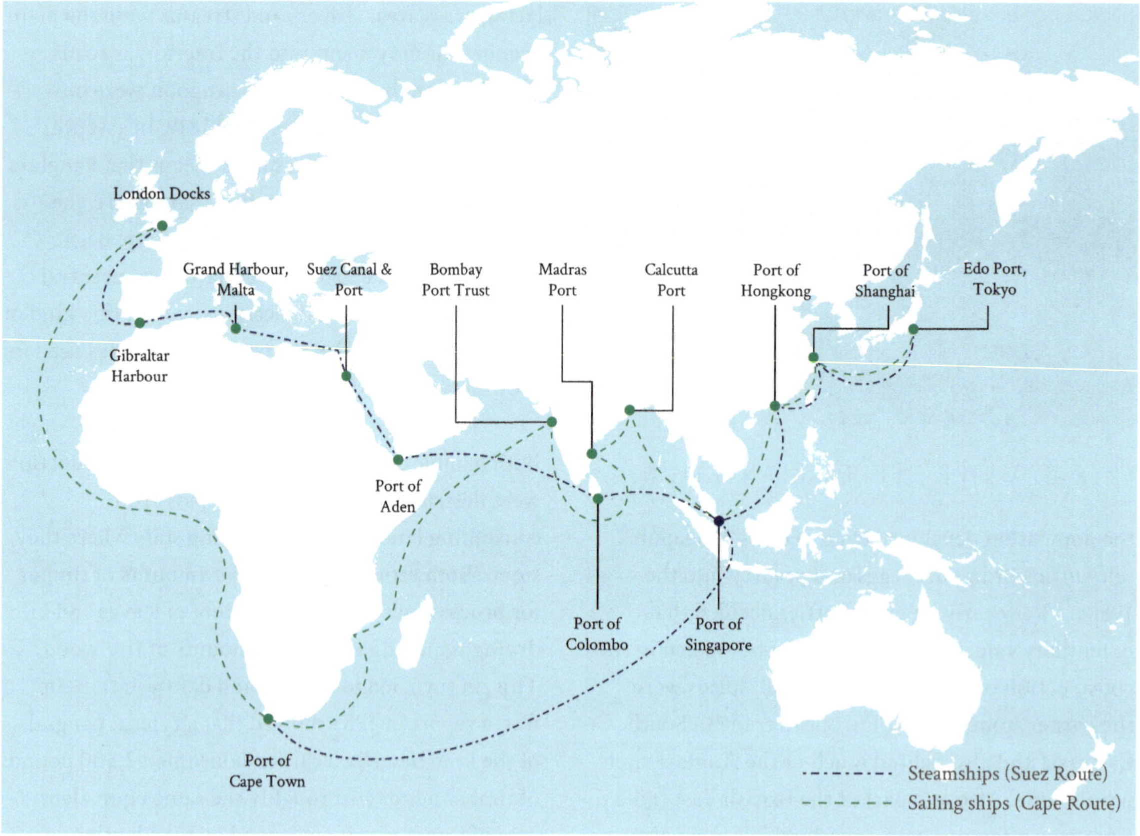

13. MAP OF SEA ROUTES TO SINGAPORE.

The second half of the nineteenth century, saw many plantations moved off-shore to Johor but the Kangchu system of settlement continued with rivers remaining the principal channel of communication.[16] By the 1840s the Europeans had ventured to Bukit Timah, the highest point on the Island and encountered the deforestation of Chinse agricultural practices. By the late 1870s the government of the Straits Settlements decided to look into the state of natural forests with the issuance of McNair report in 1879 and something of a wake-up call for government. Following that, Nathanial Cantley was commissioned to prepare a report on the Crown Forests of the colony which came out in 1883 as a scathing report on government efforts to conserve forested land. With the establishment of a Forestry Department in 1884, steps began to be made to implement recommendations to conserve forests. Generally they were divided into three categories: town reserves, coastal reserves, and interior reserves, totaling some 1,260 hectare; 920 hectare and 2,400 hectares respectively in area. This represented only 8 percent of the total area of around 540,000 hectares of land at the time. In other words, after some 50 years of gambier and other forms of cultivation over 90 percent of the forested areas of Singapore had disappeared.

By the last quarter of the nineteenth century, prosperity began to accrue to Singapore, although not in agricultural production. In 1875, for instance, Singapore had become South East Asia's major entrepôt with trade increasing something like eight times in five years. Again strategic location played a role with the Suez Canal opening in 1860 and the need for important ports for bunkering in a far reaching chain that stretched from China to Singapore to Bombay, through the Suez Canal onto Europe.[17] There were, however, four other factors behind Singapore's success apart from location. First was the style of entrepôt trade with private enterprise at its core and a cooperative engagement between government and the private sector, but with no customs controls and cohongs in between. Second, there was the presence of free trade. The monopoly of the British East India Company was broken in 1833 making Singapore very attractive to both Western and Asian traders. Third, there was the presence of a sufficient hinterland, especially post 1874 when the British gained control over the Malay States and the Straits Settlements became a Crown Colony in 1867. Finally, there was Pax Britannica in which British naval power ensured peace and prosperity keeping the sea lanes open. Indeed between 1863 and 1926 there was a 20 times increase in trade, the most substantial in Singapore's history until the past 25 years. Population pressures pushed town development into its suburbs, especially along internal roads such as Orchard Road. Migration occurred within established sections of Singapore with an ethnic remixing phenomenon, where decentralized areas became more mixed except for the further out one went into the country side. At the close of the nineteenth century the population rose to 226,842 inhabitants with 75 percent Chinese, 14 percent Malay and 9 percent Indian. There were relatively few Europeans. Also at the close of the nineteenth century Singapore became one of the world's largest sea ports, ranked around seventh behind Liverpool.

c. SECURING PROSPERITY

Around 1900, two events secured the continued prosperity of Singapore. One was the invention of the automobile and the need for rubber products produced mainly in the Malay hinterland. The second was food perseveration and innovation and the canning of goods using tin, also from Malay alluvial sources. Although not that sharply defined, the period from 1900 to the 1930s saw a productive intermixing of three intertwined phenomena. The first was booming population growth with an almost doubling from 200,000 inhabitants in 1895 to 398,000 inhabitants in 1920. This was followed by only a somewhat lesser growth between 1920 and 1940 of 398,000 to 755,000 inhabitants or thereabouts.[18] The second tied to the first was immigration and more important, public expenditure on municipal services and improvements. With the population effectively doubling between 1920 and 1930 most came by way of Chinese immigration to Singapore in probably the highest proportion in Singapore's history, at least up until that point. Many were escaping the turmoil in China at the hands of warlords on the heels of the Republic of China of 1911-12. As elsewhere, there was also a change in the laisse-faire attitudes on the part of the British, now replaced by more of a concern for social equity and social progress. In Britain the flagrance of the Victorian period could no longer be supported economically. Industrialization and companion urbanization led to counter-productive squalor and disease.[19] All of this gave rise to reform movements and concern for social causes, particularly among informed elites. The third phenomenon, again as elsewhere, was the emergence of civil society, albeit along ethnic lines but also reflected in Singapore's urban fabric. The Chinese, for instance, formed their United League in 1900 and 1910; the Malays the Singapore Malay Union in 1926; and the Indians the Singapore Indian Association in 1923.[20] Each group was pre-occupied in different ways with different issues, but local politics became very much the domain of the Straits Chinese and British attitudes were both harsh to nationalists and subversives.

14. SINGAPORE MAP, 1911

Kilometers
0 2.5 5 10

15. A TROLLEY CAR ON GEYLANG ROAD

One conspicuous upshot of this period was a concern for a widening gap between' haves' and 'have-nots, as well as improved living conditions and public works. Squalor and social divisions were reacted to through regulatory controls and expenditures on public services, public health and public safety. More specifically there was enactment of a 'light and air' regulation in 1896 and the examination of housing conditions published as the Simpson report of 1897, that eventually led to a Housing Commission or Trust to deal with squalid housing and with legislation allowing the tearing down of slums. By 1913 there was a government ordinance regulating sewer and street lighting similar to elsewhere in Britain, the United States and Europe. Certainly by the 1920s public provision of infrastructure and modern elements of planning had arrived in Singapore, including transport, public parks and roads. Again as elsewhere, more affluent suburbs became better served than poorer inner city areas. Reservoir construction that had been more or less continuous since 1867, mainly due to population growth, with the MacRitchie Reservoir, named for the engineer who designed and built it.[21] The Kallang Reservoir followed in 1911 and the Sungei Seletar Reservoir followed 1922. Effectively in 1900 the entire catchment of the Kallang and Seletar Rivers above the dam walls became the Municipal Catchment, also protecting secondary and primary forest within that area. This was later gazetted as a Nature Reserve in 1951. In retrospect construction of these reservoirs was seen, by some, as both a blessing and a curse. Of course there was some abundance at the time of clean water and vegetation conservation, but also the loss of fresh water marsh forest.

Returning to rubber production, Singapore though a minor producer, had major impacts on research into para-rubber trees at the botanic gardens. There the work of the botanist Henry Nicolas Ridley was instrumental in promoting rubber trees in the Malay Peninsula and making the industry economically viable.[22] In Singapore itself there was some production in the form of several large plantations, on the order of 8,000 hectares or so in size. The Great Depression, however, at the end of the 1920s and into the 1930s adversely affected rubber and latex production with prices falling precipitously. Over time, the declining rubber plantations were replaced by intermixed vegetable farms well into the Japanese occupation. In addition to rubber, tapioca and Liberia coffee were experimented with as cash crops, particularly by Leopold Chasseriau on his 485 hectare estate adjacent to the central catchment area. These experiments had, however, no lasting impact.

Between 1819 and 1945 Chinese Singaporean farming was effectively locked into the British system of distribution of economic development.[23] Small-scale local family farms increased agricultural output to meet demands of rising populations, producing cash crops like gambier, pepper and nutmeg. With withdrawal of the East India Company in 1834, some former employees also farmed plantation agriculture for profit. Chinese businessmen often supplied the system with labor via a 'Credit Ticket System' of labor recruitment. Under this punitive arrangement, Chinese passage was arranged and paid for by working at least a year at reduced wages. In addition, a proportion of the wages was paid for in-kind with food, clothes and opium. The resulting cheap labor was thus exploited for agriculture with the small-family farms excluded from these processes and receiving little government support. Many cheap laborers also lived in the over-crowded areas of Chinatown and other locales, exacerbating the squalor and poor quality of life in the city mentioned earlier. In 1905 a cholera epidemic claimed 759 victims, mostly Chinese. Unlike the plantations that were inflexible with their crops, plagued by poor and exhausted soils and encountering plant disease, the small-scale farms of the Chinese agriculturists proved to be flexible, low in overhead and highly productive of food. Indeed, in 1945 family farms were the most important agricultural activities on island, despite also producing pollution problems.

To briefly summarize the state of affairs before World War II, the early colonial administration of Singapore engaged in little land management. Population expanded appreciably. There was a minor boom in rubber latex production using Ridley's herring-bone method of extraction. Catchment areas were developed along the reservoirs. However, the original dryland primary forest shrank from 410 square kilometers to 2.01 square kilometers; the fresh water swamp forest declined from 74 hectares to a mere 1.33 and the mangrove forest from 87 ha to 5.7 square kilometers. Overall the original primeval forest of 571 square kilometers diminished down to 9.04 square kilometers, or a mere 1.58 percent of its original scale. Mostly this was due to unproductive agricultural production and misguided land management. [24]

16.MAP OF SINGAPORE'S PRIMEVAL FORESTS

17. A DEFORESTED PLANTATION IN 19TH CENTURY SINGAPORE

d.

WAR-TIME INTERLUDE

At the time of Japan's military campaign to create its Greater East Asia Co-Prosperity Sphere defenses at Singapore, both at Changi and Jurong, were oriented towards seaward attack. Instead, Lieutenant-General Tomoyuki Yamashita's army swept down the Malay Peninsula crossing easily into Singapore Island. After valiant resistance by Commonwealth and local forces, Singapore fell to the Japanese in February of 1942, also opening the way to conquest of the Netherlands' East Indies. The Syonan years ('light of the South'), as the Japanese occupation has been called, was a period of considerable confusion for the people and defenders of Singapore.[25] For more than a century the security of Singapore in British hands was taken for granted. Under the Japanese the city with a population on the order 800,000 was ran by a municipal administration or *tokubetsui-si* but under the military administration or *gunseikan-bu* with the Kemppeitai, or military police, holding sway. What took place was a 'reign of terror' during the first months of occupation, if not longer, involving purges, torture and massacres. Generally the Chinese population was brutally treated, partly because of gallant and persistent Chinese resistance to the Japanese attackers. The Malays were left somewhat more alone and the Indians were invited to join forces with Japan, which some did. All the Westerners were interned in concentration camps.[26] This all occurred despite earlier Japanese expressions of Asian brotherhood. In time the Japanese brutally dashed any hopes the conquerors may have had in gaining support of the local Singaporean population.

18. JAPANESE OCCUPATION OF SINGAPORE.

Apart from trying to coerce the population to unify as Asians, the Japanese set in place policies to sweep away any of the earlier colonial super structure. Instead Syonan was to become a self-sufficient state with regard to its industry, communications, commerce and finance all harnessed to the war effort.[27] This was a significant change for the free-trading and outward-oriented entrepôt of earlier times. Large Japanese enterprises like Mitsui and Mitsubishi were given control of important branches of the Malayan economy. Import substitution was practiced, albeit in rudimentary form. Pineapple fiber, for instance, was substituted for rope and string; paper was made from bamboo; and grease and lubricants made from palm oil. By the middle of 1942 there was a large influx of Japanese businessmen to Singapore and the first Singaporean built steam ship was launched in November of the same year.

Conversion to Japanese military script, by way of currency, hastened rampant inflation. Singapore was further enfeebled economically through a pernicious policy of income extraction, mainly at the expense of the Chinese population, that was to be gifted to Japan to help them in their war effort. By and large Japanese possessions were seen to be self-financing with regards to services. Food also became in short supply, especially with regards to staples such as rice and also more generally due to the overall shift to self-sufficiency and away from free trade with the outside world, one of Singapore's lasting hall marks. No more benignly the Japanese set about to alter the Singaporean education system, stressing loyalty and national consciousness with strong accents on technical and vocational instruction. With regard to natural and real assets, the reservoirs were preserved, vegetable gardens were encouraged in-lieu of plantations and cash crops.[28] The Japanese were also persuaded to preserve the botanic gardens and the famous Raffles Museum collection, with Kwan Koriba, a Japanese botanist in control.

The war-time industrialization campaign, however, proved to be superficial and ephemeral, although also demonstrating Singaporean ingenuity and capacity for further industrialization.[29] Nevertheless, as one author put it: "Instead of Asian brotherhood the Japanese brought cruelty and tyranny". They also conveyed racial arrogance and a feeling of superiority that also undid many of their plans. Little came of Japanese schemes for physical redevelopment of Singapore other than clearing buildings and basically rehabilitating services from prior to the occupation. One conspicuous exception was the construction in 1943 of an airport at Changi using prisoners of war labor that was completed in 1945, becoming the forerunner or today's vast international airport. The year 1943 was also something of a turning point in Japans fortunes of war. At much the same time the nascent Indian independence movement, tacitly if not modestly supported by Japan, began to gain traction. By November of 1944 the Americans began air raids over Singapore Harbor though careful to not inflict damage on non-Japanese sites. In August of 1945 the Japanese surrendered to the Allied Forces and in September Singapore was liberated. Rule was handed back to the British, although continued British hegemony was in doubt. In the months immediately following the collapse of British power in Malaya, the Colonial Office began drawing up schemes for radical post-war reorganization. One of its first proposals grouped the Malay States; Straits States; North Borneo; Sarawak and Brunei into a single union, with Singapore as its natural centre of trade and communication. This was abandoned as too complicated and out of the feeling that London did not want to jeopardies Malayan Union, important to them against communist insurgents, by trying to include Singapore. Instead it saw Singapore as a vital free port, defensive base, and a way of assuaging Malay strong fears of Singaporean domination should they be included in the Union.

e.

POST-COLONIAL DEVELOPMENTS

Under its avowed commitment to decolonize, while ensuring a non-communist take-over of its possessions, Britain pursued a policy of tactical convenience and separation of Singapore from other states in the Malay Federation. The United Malay National Organization, or UMNO, was founded in 1946 under Tunku Abdul Rahman and took a strong position against the union of Singapore in what became Malaysia.[30] The origins for this were ethnic, cultural and also, as mentioned earlier, the fear of Singaporean domination. On the Singaporean side, in 1949 Lee Kuan Yew and some of his fellow students from Cambridge set up the Malay Forum and decided to push for an independent, socialist Malaya, but included Singapore which they saw as being too small to survive on its own. Later on in 1954 Lee formed the People's Action Party (PAP) together with allies from the Malay Communist Party, taking an anti-colonial line which the communist supported more strongly. Then in 1959 during the general election for an autonomously governed Singapore, Lee and the PAP won with Socialist although not Communist support. In 1963, then-Prime Minister Lee Kuan Yew signed the Malaysia Agreement in London. This resulted in the formation of the Federation of Malaysia consisting of Singapore, Malaya, Sarawak and North Borneo (Sabah). However, in August of 1965 via the Singapore Amendment to the Malay Federation, Singapore was summarily ejected. Suddenly Lee and the PAP were faced with a nation building process despite their wishes and prior commitments. Quoting Lee Kuan Yew at the time, "We must survive... Just as a river loops and bends before it reaches the sea, so the history of a people takes many loops and bends before it reaches its destiny... Everybody will have their place, equal language, culture and religion... we could not achieve multiracialism and integration in Malaysia. But we will achieve it in Singapore".[31]

When the leadership looked around Singapore what did they find? In short, they found a fragmented society with substantial socio-economic and dependency problems. First there was rising unemployment, as much as 17 percent to 20 percent and declining free trade.[32] Second there was exploding population growth with refugees from Malaya shifting from re-federation, among others. The actual rise was from about 1 million people in 1950 to over 2 million by 1970. Third, there were appalling living conditions for many with spot densities of 2000 people per hectare and less than 10 square meters of livable space standard per person. Much of this was due to massive over-crowding. There were also some 300,000 people in squatter settlements in addition to around 250,000 people living in slums. Segregation of population also led to race riots in the 1960s. Then there was substantially deteriorated infrastructure due to neglect in the wartime years and finally there were shaky situations with neighboring countries, like the conflicts in Indonesia of 1965 and the water dependency on Malaysia through the water agreements of 1961 and 1962. In these last regards, Singaporeans also remembered well a time during the Japanese invasion when water was severely constrained due to damaged supply pipes, placing them in dire strait. The 1961 agreement signed by the Singapore City Council with the State of Johor in Malaysia, lasted until 2011, whereas the 1962 agreement extended the supply up to 250,000,000 gallons a day from Johor until 2061[33].

19. SQUATTER SETTLEMENTS IN SINGAPORE

Economic prosperity was to be secured by capitalizing on local assets. These included a relatively low cost, well-educated local labor force, dating back into the colonial period. Also, Singapore was to make itself open to trade as in the past and to encourage investment. It was also to move from a dependence on entrepôt trade in this regard to modern industrialization. Moving forward, Singapore engaged in an effective form of 'collective consumption' led by the government making Singapore an attractive place for foreign investment and manufacture.[34] Here there was to be an insistence on political stability and a clean bureaucracy. The English-speaking legacy of the past was also to be continued lowering potential friction with potential outsiders from the United States, for instance, and from Europe. Then there were plans to create sophisticated infrastructure and

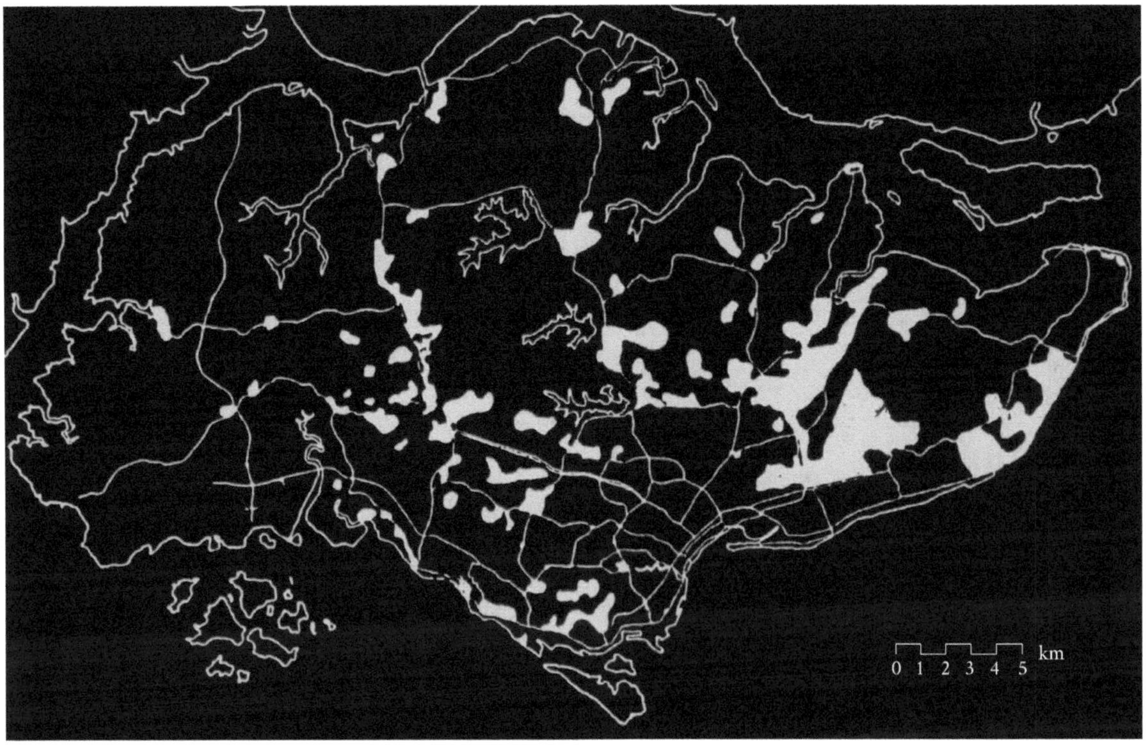

20. MAP OF SINGAPORE'S SQUATTER SETTLEMENTS

other basic improvements that would foster economic growth, capital accumulation and attract foreign investment. This involved three basic components. First, there was to be provision of public housing to help reduce and stabilize living costs and thereby enable the provision of cheaper labor. Then there were plans to build industrial infrastructure to offset the need for extensive foreign investment and, finally, to provide ready facility and turn-key projects for foreign enterprises such as those in the Jurong Industrial Park. Throughout, use was to be made of institutional boards and other British legacies. The Housing and Development Board (HDB) was established in 1960 as one of the super boards and then the Economic Development Board was begun in 1962. The Jurong Town Corporation, for example, was paired with the Economic Development Board and the Urban Redevelopment Authority (URA) was placed in relationship to the Housing and Development Board as an authority with both regulatory and developmental power, similar to Boston's Redevelopment Authority in the United States.

To both guide and orchestrate urban development activities and to promote orderly urban growth. Master-planning exercises were conducted across the island in several steps. First, British ideas on future development of Singapore conceived in the waning years of colonial rule were contained in the 1958 Masterplan. Then, E E. Lorange, a Norwegian planner, was hired with United Nation's assistance to produce a revised plan, followed in 1963 by a further United Nation's sponsored team of Koenigsberger, Abrams and Kobe that led to a so-called 'Ring City Plan', incorporating a more or less circular transportation armature around the central catchment area of the island and from which urban development could be organized[35]. This effort was extended in 1965 into a competition for consultation which was

21. A NIGHT CART IN OPERATION

won by the Australian firm of Crooks, Michael, Peacock, Stewart Pty, Ltd. and led by Henry Wardlaw an Australian from Sydney. This exercise resulted in the 1971 Concept Plan, which incorporated the earlier 'Ring City Plan' idea and elaborated on it to produce a plan for Singapore capable of accommodating a population of four million inhabitants by 1992.[36] In addition to the transportation ring around the central catchment area, by way of two north-south corridors, sites for new communities and housing estates were located along these corridors, including both expressways and lines of mass rail transit. Heavy industries were planned at the western end of the island in Jurong, and Singapore's international airport was planned at the eastern end of the island in Changi, also with some industry. Both locations avoided conflict with existing and planned urban areas and both were also to be integrated with the island-wide mass transit system. Overall, the plan provided for a heavy belt of urban development along the south coast, with intermediate intensities around dispersed centres.

A master-planning process was established whereby 'concept plans' would be reviewed every ten years, but with mandatory quinquennial reviews of master plans set within the successive 'concept plans'.[37] In effect, this has yielded 'concept plans' in 1971, 1991, 2001, and 2011, with meso-scale master plans in 1965, 1970, 1975, and 1985. Before 1998, the Master Plan served as a "blueprint" document that was only updated periodically to capture developments that already existed, or had already been approved. The Master Plan did not actively communicate future planning intentions and permissible uses, and planners were instead guided by confidential "drawer" plans when assessing development applications. The "Development Guide Plans" prepared in the 1990s laid out future planning intentions for each planning area in a transparent and systematic manner. The plans were completed by 1998. This process contributed to the development of the new Master Plan, which was gazetted, detailing future plans and permissible future developments on each parcel. Once the planners had a good base of Master Plan, there was no longer a need to develop similar 'Development Guide Plans', though the process of it are incorporated into each Master Plan review.

With regard to water resources, after 1965, Singapore intensified its efforts to diversify the sources of water supply rather than depending mainly on the water from Johor. In 1971 a Water Planning Unit was created within the Prime Minister's Office to study the scope and feasibility of new conventional sources of supply, as well as unconventional sources involving water re-use and desalination. This study resulted in the first Water Master Plan of 1992, outlining strategies for local resources involving more local catchment, NEWater recycled water and desalinated water. Initially, the Water Planning Unit sought the expertise of Tahal Consultants from Israel. However, the water resource conditions in Israel were very different from those in Singapore and so the Water Planning Unit struck out more independently on its own master plan with a 20-year horizon until 1998.[38] They also acknowledged that, over time, any specific single water conserving

22. A TURNKEY FACTORY IN THE JURONG INDUSTRIAL PARK

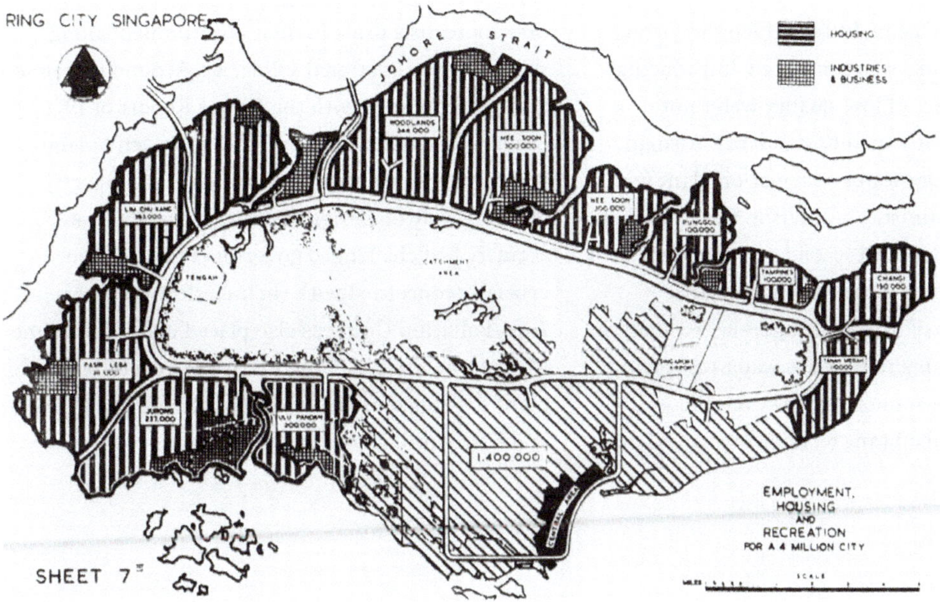

23. THE KOEINGSBERGER RING CONCEPT PLAN OF 1963, WHICH WAS EVENTUALLY INCOPORATED INTO THE LATER 1971 CONCEPT PLAN

industry could upset demand forecasts both widely and quickly. The supply system options in their 1972 plan included other surface water sources including reservoirs at Bedok, Kranji-Pandan Dam and within the western and eastern catchments. They also recognized the need for water supply co-existing with other adjacent land uses, meaning urban stream water generally, as well as estuary barrage sources of supply and damming of rivers. Ground water proved to have poor potential at the time and so serious consideration began to be focused on recycled or reused water. In this regard treated effluent as a source of potable water supply was not advocated. But particulate objects might be removed, if the treated effluent was first diluted in a large impounding reservoir affording long retention periods and undergoing self-purifying processes prior to withdrawal for treatment needs. Desalination, by contrast, was considered a reliable source not subject to weather fluctuations, although costly still to bring on-line fully. In 1966 the Jurong Industrial Waterworks was completed, introducing an inexpensive source of low-quality water not requiring potable water quality standards. Also, in 1974 an advanced pilot water reclamation plant was established by the Ministry of Environment and then-Public Utilities Board, with secondary treated effluent undergoing reverse osmosis and other advanced treatment processes of iron exchange, electro-dialysis and ammonia stripping to produce water complying with drinking water standards. However, the membrane in this membrane technology proved to be expensive and not very reliable. The plant closed after fourteen weeks. Nevertheless, the die, so to speak, was cast for NEWater to come on-line. Then on the Malaysian front, the 1962 water agreement was supplemented by the 1990 water agreement, scheduled to expire in 2061.[39]

In 1963, Prime Minister Lee Kuan Yew put forward the idea of Singapore being a 'clean' and 'green' 'garden city', despite high-density development, limited land space and other resource constraints.[40] He also started the 'tree planting program' that was to have fairly immediate effect on Singapore's urban landscape. One compelling reason for the 'clean and green' program was the need to collect as much of the 241.3 centimeters per year in rainfall as was possible. In this regard the 1963 drought was something of a wake-up call for further action. Another compelling reason was to make Singapore an attractive city, especially for direct foreign investment and to give Singaporeans a pride in their environment along with a sense of national solidarity. Around this time and in accordance with the Cleary Report of 1970, Singapore adopted an integrated approach to land development, mitigating environmental impacts, building only on designated sites. In effect, water security had challenged government to become crucially concerned with their environment. An Anti-Pollution Unit was also placed under the Prime Minister's Office, helping Lee Kuan Yew to realize his plans for Singapore to quickly become a clean, green and first-world city-state.[41]

24. PRIME MINISTER LEE KUAN YEW PLANTING A TREE

endnotes

1. Turnbull, C.M. 1977. *A History of Singapore,* 1819-1975.First Edition, (Singapore: National University of Singapore), p.2.

2. Turnbull, p.1.

3. Sathiamurthy, E. and H. K. Voris.2006. "Maps of Holocene Sea Level Transgression and Submerged Lakes on the Sunda Shelf" *The Natural History Journal of Chulalongkorn University,* p.1f.

4. Turnbull, p.3.

5. Paragraph based on Turnbull, p.3-4.

6. Chew, Ernest C. T. and Edwin Lee. Eds. *A History of Singapore* (New York: Oxford University Press), pp..116-117.

7. Turnbull, p.4.

8. Paragraph based in part on Turnbull, pp.4-5.

9. Paragraph based in part on Perry, Martin; Lily Kong and Brenda Yeoh. 1997. *Singapore: A Developmental City-State* (New York: John Wiley and Sons), p.23 and p27.

10. Turnbull, p.11.

11. See location and dimension on the accompanying plan.

12. O'Dempsey, Tony. 2014. "Singapore's Changing Landscape Since c. 1800", in Timothy P. Barnard ed. *Nature Contained: Environmental Histories of Singapore* (Singapore: NUS Press), p.18.

13. Barnard, Timothy P. 2014. "Introduction", in Timothy P. Barnard, ed. *Nature Contained: Environmental Histories of Singapore* (Singapore: NUS Press), P.3.

14. Paragraph based in part on O'Dempsey,p. 21 and p.22.

15. Paragraph based in part on O'Dempsey, p.23 and 0.28.

16. Paragraph based in part on O'Dempsey, p.28, p.32 and p.32f.

17. Paragraph based on Turnbull, p.73, p.95 and p.97.

18. See graph of population statistics in Chapter 1.

19. Turnbull, p138f.

20. Turnbull, p140f.

21. Paragraph based in part on O'Dempsey, p.38, p.39 and p40.

22. Paragraph based in part on O'Dempsey, p.35 and p.42f.

23. Paragraph based in part on O'Dempsey, p.29 and p.47.

24. Paragraph based in part on O'Dempsey, pp.20-30

25. Perry, p.46.

26. Turnbull, p.190f.

27. Paragraph based in part on Turnbull, p198, p.200, p201 and p.202.

28. Taylor, Nigel P. 2014. "The Environmental Relevance of Singapore", in Timothy P. Barnard, ed. *Nature Contained: Environmental Histories of Singapore* (Singapore: NUS Press), p.129f.

29. Paragraph based in part on Turnbull, p.201, p.205, p.206, p.211, p.216, p.217 and p.218.

30. Partially based on Hack, K., Margolin, J.L., and Delaye, K. (2010). *Singapore from Temasek to the 21st Century: reinventing the Global City* (Singapore: NUS Press) Chapter 10, pp.243-291

31. Transcript of a press conference given by the Prime Minister of Singapore, Mr. Lee Kuan Yew at Broadcasting House, Singapore at 1200 hours on Monday 9 August, 1965.

32. Perry, p.48f

33. Tortajada, Celia; Yugal Joshi and Asit K. Biswas. 2013. *The Singapore Water Story: Sustainable Development in an Urban City-State* (New York: Routledge), p.14.

34. Castells, Manuel, Lee Goh and R.Y. Kwok. 1991. *The Shek Kip Mei Syndrome: Economic Development and Public Housing in Hong Kong and Singapore* (Los Angeles: Sage Publications, Ltd.)

35. Perry, p. 192.

36. Interview with Dr .Liu Thai Ker on 10 August, 2017.

37. Perry, p. 192f.

38. Tortajada et al., 2013. p15-16.

39. Paragraph based in part on Tortajada, p.15, p.16, p.17 and pp.24-25.

40. Paragraph based in part on Tortajada, p35, p.39 and 46.

41. Paragraph based in part on Tortajada, pp.64-65.

chapter

03 VISIONS OF CLEAN AND GREEN

P. G. Rowe and L. Hee, *A City in Blue and Green*,
https://doi.org/10.1007/978-981-13-9597-0_3

Against this background, Lee Kuan Yew at the head of the People's Action Party (PAP), now in the seat of power in the new Republic of Singapore, prosecuted his insight that Singapore should become a clean and green city. This was motivated, as mentioned earlier, by putting the right face on the attraction of outside foreign investment and tourism upon which Singapore was to depend. He also saw 'clean greening' to be a tangible way of addressing inequality throughout the island and, finally, as an opportunity to create an identity for residents and a sense of national pride. As elsewhere, this underlying narrative or metaphor for island-city development went through subsequent transformations. The 'Garden City' idea, for instance, moved on to be a 'City in a Garden' and then towards the end of the twentieth-century, 'a Tropical City of Excellence' envisioned by the URA with strong connotations of its vegetative and 'geographical 'green-blue' aspect. More recently, more of a convergence into a 'city in nature' is emerging, with greater emphasis strongly on the identity aspect of an altogether tropical landscape in which living takes place befitting Singapore's geographic and otherwise natural location. This latest 'turn' on 'garden' and 'city' also argues for greater complementarity of green and blue aspects of Singapore's intrinsic environment. It involves efforts to intensify the planting of more native and more diversified plant species; and through multi-layered heights of plants to emulate the forest structure and to create more ecosystems providing niches for different fauna species. The plantings are thus carefully curated in the beginning but will be left to grow and evolve naturally to produce a naturalistic rather than garden-like setting. It may also provide the rest of us with a clearer view into the successful occupation of 'hyperobjectivity' by beginning to invert the usual dominance of constructed over 'natural' environments in city making. Of course, Singapore is not alone in this strong metaphorical and influential presence of 'natural and cultivated' landscape qualities with those of a more constructed and technological kind, nor of the transformation of such juxtapositions over time. American 'pastoralism', for instance, giving way to the 'machine in the garden' is another such instance. Also, various rounds of '*citta*' in '*compagna*' in the Italian ecumen is another along with other European schemes. Indeed, it seems as if such poetics of occupied and environmental space are leit motifs of civilization. The key point here, though, is that what Singapore is beginning to tackle, so to speak, is unique and of its own making.

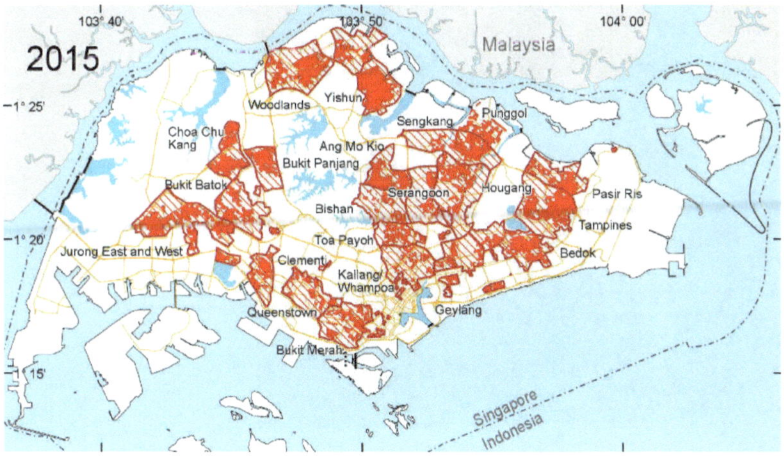

25. HOUSING AND DEVELOPMENT BOARD ESTATES IN THE CONTEXT OF URBANIZATION, C.2015

a.

ONSET AND RISE OF 'CLEAN AND GREEN'

In a recent memoir, Lee Kuan Yew broached the subject of Singapore being 'clean and green' in the following manner. "After independence, I searched for some dramatic way to distinguish ourselves from other third-world countries. I settled for a clean ad green Singapore".[1] He went on further to say, "one aim of my strategy was to make Singapore into an oasis in Southeast Asia, for if we had first-world standards, the businessmen and tourists would make us a base for their businesses and tours of the region." [2] Moreover, "if we want to be a first-world oasis, we must produce first-class conditions, not just the environment but facilities, health standards, services, communications and security."[3] In short the stress on 'clean and green' was both a pragmatic reckoning with need and a metaphor for how he wanted the new island-state to be seen. Although ostensibly about a literal 'greening' of the state, it was to be more inclusive by attaching to almost every walk of life and to define the manner of interaction with the outside world, as well as to Singaporean senses of themselves and a more equitable distribution of this amenity benefit, As lee put it, " if we did not create a society which is clean throughout the island, I believed then and believe now, we have two classes of people: the upper-class, upper-middle class and even the middle class with gracious surroundings; and the lower-middle and working class, in poor conditions."[4]

26. TREE PLANTING IN SINGAPORE

Apart from formal programs, a number of clean-up campaigns were initiated during the 1960s and into the early years of the city-state's development. Littering in various forms, for instance, came under scrutiny and regulation.[5] This began in October of 1968 with the month-long 'Keep Singapore Clean Campaign' aimed to discourage public littering as a part of a longer program that included changes to public health laws, development of sewage systems, disease control and even the relocation and licensing of itinerant hawkers.[6] The latter had proven to be a nuisance because of the dirt and stench of rotting food, as well as obstruction of public rights-of-way. Anti-spitting and anti-chewing gum campaigns also ensued in attempts to clean up the place and encourage more civilized behavior on the part of citizens. Taxi drivers without licenses and adequate insurance and who drove rented junk cars were also cracked down upon, a situation which was not resolved satisfactorily until 1971. Allegedly one morning in 1964, Lee Kuan Yew looking over the Padang from city hall saw several cows grazing upon The Esplanade. Shortly thereafter all stray animals were seized or slaughtered.[7] Part of the greening process also involved bringing Singapore more closely in harmony with its environment and striking the right balance between industrialization and environment. In these regards stricter standards were insisted upon in large-scale operations, such as the Jurong petrochemical refinery with Sumitomo.[8] Upon returning to Singapore from a trip to Boston, Lee Kuan Yew noticed that the roadside trees were covered with grime from cars and trucks, unlike in the American city. He also discovered this was because of annual inspections required of all vehicles in the United States that effectively limited pollution. Soon the Singapore government introduced inspections and fines on polluting vehicles. Tree planting started in 1963 and by 1971 a regular tree-planting day program became initiated annually, usually taking place in November as one of the best times of year to undertake such activity. This was followed in 1976 by a campaign to plant hedges, creepers and foliage along fencing structures and in association with the city's concrete infrastructure like retaining walls, overhead bridges and flyovers.[9] In addition in the 1980s color was added to the palette of trees to compensate for a city that seemed to be too green, resulting in Bougainvilleas, Hibiscus and Ixoia shrubs being planted throughout the island.

Efforts to clean up the island also resulted in the closing down of numerous small-scale often family-run farms. By 1982 Singapore was self-sufficient in the production of pork, eggs and chicken, as noted earlier, with a total value of agricultural production in 1968 standing at around US$285 million.[10] Then in 1984 the Division of Primary Production was tasked to improve the efficiency of agriculture which they tried to do by advocating the phasing out of smaller farms. These businesses were also seen at the time as being backward and very polluting, especially the pig farms. The planned transformation of farmlands into agro-technology parks never really transpired and today agriculture accounts for only 0.2 percent of total GDP.[11] Certainly by the mid-1980s, if not before, Singapore was cleaner and greener than at the time of independence in 1965. Apart from moving closer to becoming a first-world city within the Third World, its avowed ambition, the three other aims in becoming clean and green were becoming evident. The first was attraction of outsiders in the form of investors, business people and tourists. The second was to provide a stronger sense of pride and identity for Singaporeans. The third was to effectuate a more equitable distribution of natural amenity across the island, without disparities among neighborhoods, particularly between those that were richer and less so. It was also a place where the 'city' part of the 'garden city' formulation also came to receive attention with regard to historic conservation, in addition to the shiny new projects of the Housing and Development Board (HDB) and the Urban Redevelopment Authority (URA).

27AB. THE GREENING PEDESTRIAN BRIDGES AND TRAFFIC ISLANDS

b.

FROM 'GARDEN CITY' TO 'CITY IN NATURE'

The Garden City Movement is a method of urban planning that was introduced in 1898 by Sir Ebenezer Howard in the United Kingdom. Garden Cities were intended to be planned, self-contained communities surrounded by greenbelts and containing proportionate areas of residences, industries and agriculture. Inspired by social utopian views about living from towards the end of the nineteenth century. Howard published the book *To-morrow: A Peaceful Path to Real Reform* in 1898, which was subsequently re-issued, as *Garden Cities of To-morrow* in 1902.[12] Howard's idealized garden city would have from 30,000 to 35,000 inhabitants and be planned in a concentric pattern with open space, parks and broad radial boulevards extending out from its centre. It would also be self-sufficient and could be linked to similar communities forming satellites to a large, older urban centre. Not inappropriate for Singapore, the beginning of the Garden City Movement came at a time of reform of the overcrowding, dilapidation and squalor of large industrial towns and cities during their industrial revolutions. At the time it also represented a town-country merger and drew upon the opinion of many about the deleterious nature of cities that were seen as being one of the biggest issues to be confronted at the time. Soon the 'Town and Country Planning Association' was formed and the 'First Garden City, Ltd' chartered to build Letchworth in 1899 some 50 kilometers outside of London. This was followed by Welwyn in 1919 even closer in to central London.[13]

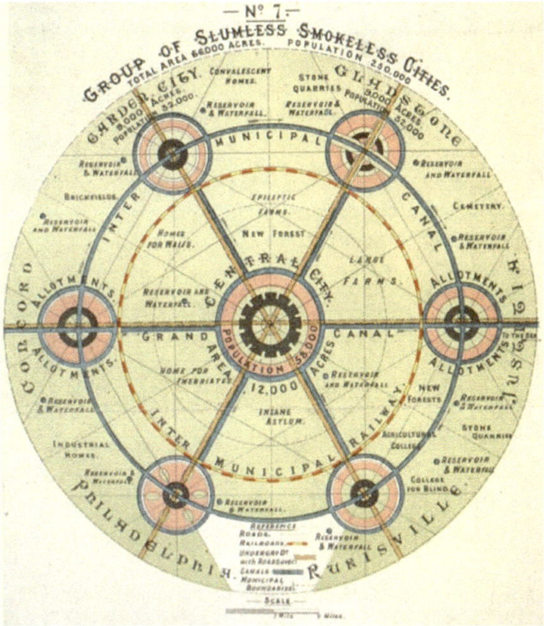

28. HOWARD'S GARDEN CITY OF TOMORROW

From the early beginnings an international movement quickly formed with adherents and proponents in most of Europe, the United States, South America, parts of Asia and in Australia. Indeed, Singapore has incorporated various facets of the Garden City concept over time, beginning with its Concept Plan of 1971. Since then building codes, land-use plans and projects have made adequate provision for greenery to become integral part of urban development. One of the exceptional aspects of Singapore as a Garden City is the density of its development in comparison to the relatively low density of many other Garden Cities. In fact, in this respect it stands apart from earlier criticism levelled at Garden Cities because of destruction of the countryside, poor movement circulation and a lack of convenience. In many places the concept also

29A. LETCHWORTH IN THE U.K.

29B. WELWYN IN THE U.K.

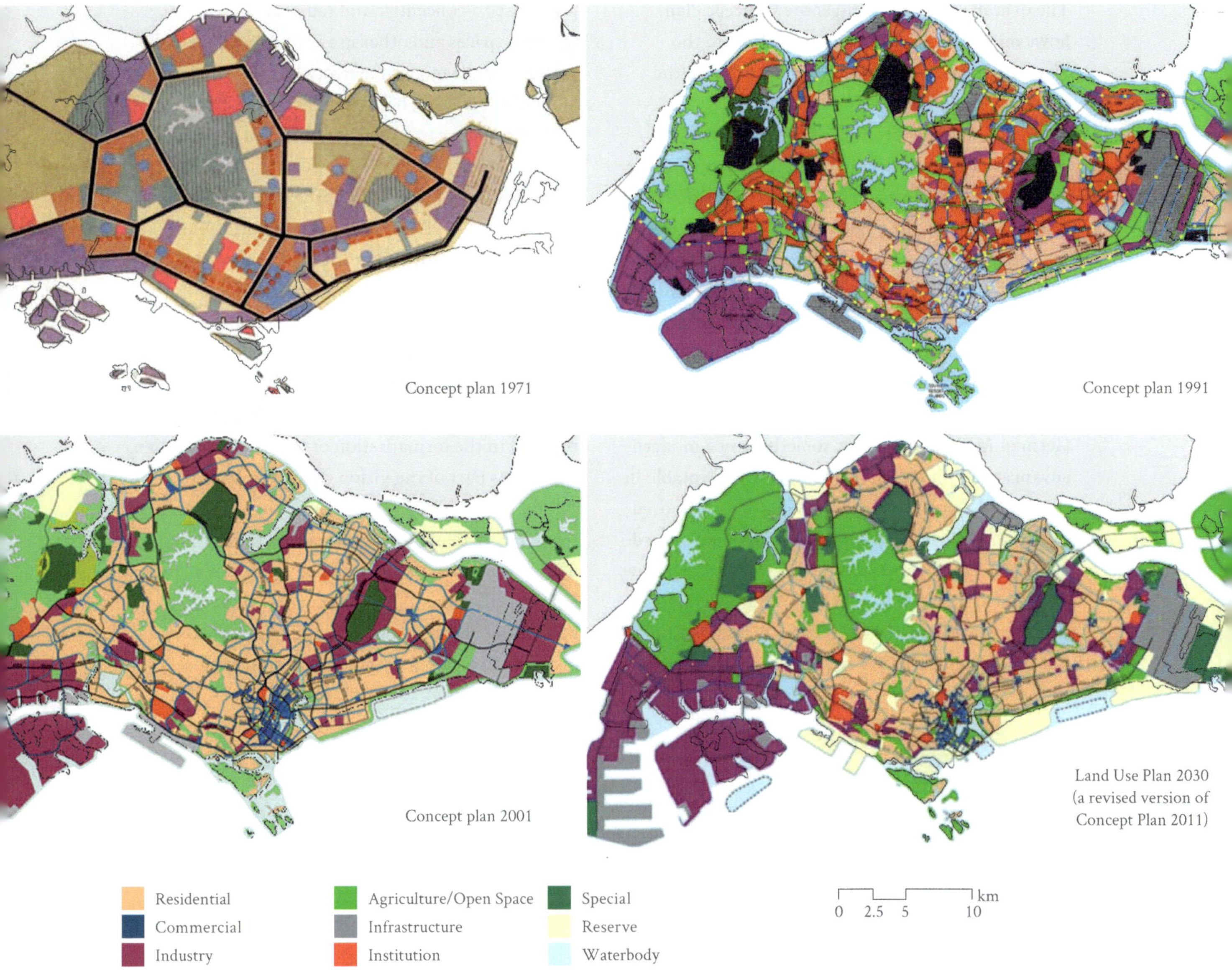

30. SINGAPORE'S STRATEGIC LONG-TERM LAND USE AND TRANSPORTATION PLANS

devolved, unfortunately, into dull dormitory towns with few redeeming features. Looking at Singapore's Concept Plans of 1971, 1991, 2001 and 2011, tropes of Howard's earlier plans can be found. First, there is the division of land into zones, particularly separating industry from the other uses. Second, there is the large catchment areas at the centre and both eastern and western sides of the island, which provide for extensive natural conservation areas, as well as public parks and recreation areas. Third, there are the major expressway and transit corridors that form the armature of connection between and among satellite communities and new towns. Fourth there is the idea of self-contained communities and in a manner similar to the Garden City ward and centre arrangement proposed by Howard in 1902.

The overall scale of the Singapore Concept Plan, however, is also close in degree and kind to the Plan for Greater Berlin of 1910. This was also the result of a competition, won by Herman Jansen, but including other prominent proposals and diagrammatic considerations of large urban areas. At the time Berlin was undergoing strong urban growth, carrying the population from around one million inhabitants in 1900 to around four million by 1910.[14] This was causing considerable overcrowding and congestion, not to mention the administrative fragmentation that had become a significant problem resulting in the unification under 'Greater Berlin'. What also loosely emerged was the model of the German Metropolitan City, something of a modern advancement on the smaller Garden City. Notable in this regard were the diagrams and layouts proposed by Bruno Möhring, Rudolph Eberstadt and Richard Petersen in their third placed scheme.[15] Möhring was an architect. Eberstadt an economist and Petersen was a traffic engineer. Their urban diagrams, similar to Howard's Garden City concentric ring depiction, juxtaposed a concentric and radial arrangement of green spaces and other uses as a developmental pattern for the modern metropolis. Residential and industrial uses, both very prolific in Berlin at the time were to be separated and an organic link between the city and its green landscape was to be established. These and other later planning developments in Berlin, even after World War I were to more clearly establish the framework for the large-scale modern urban metropolitan city of which Singapore is more recently of a similar kind.

Eventual success in tree planting and greening schemes during the early days of the Republic resulted in the formalisation of the City in a Garden concept as part of the vision of the Garden City Action Committee in 2004.[16] More explicitly this refers to urban environments where the built man-made elements appear to be located in a green landscaped garden.[17] A fundamental philosophy behind the vision was to bring people close to greenery in a city and to integrate it more closely

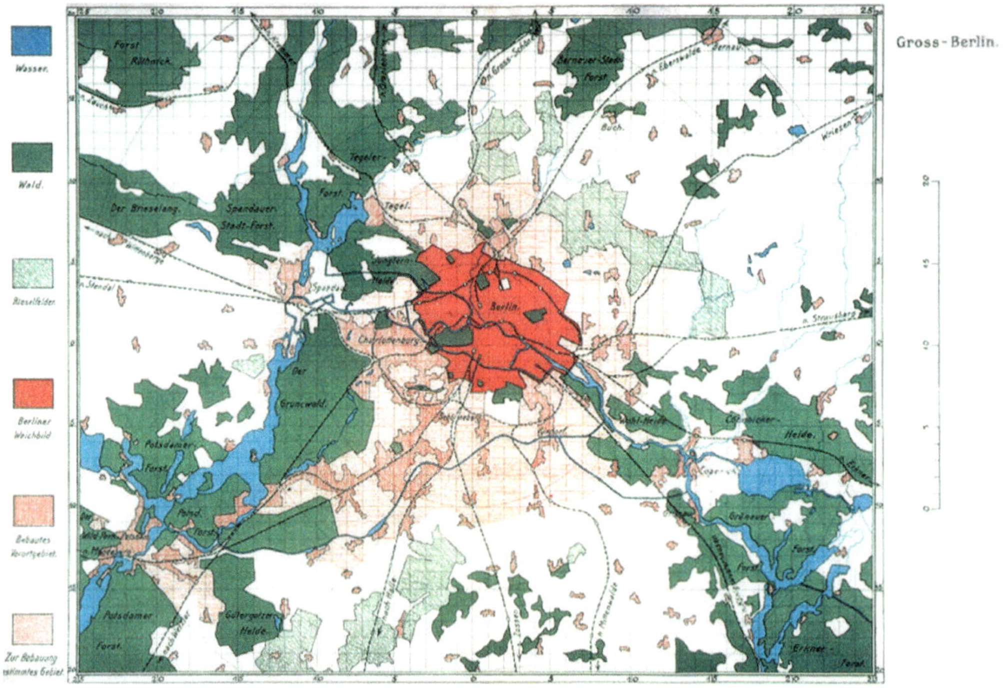

31. THE 1910 BERLIN PLAN

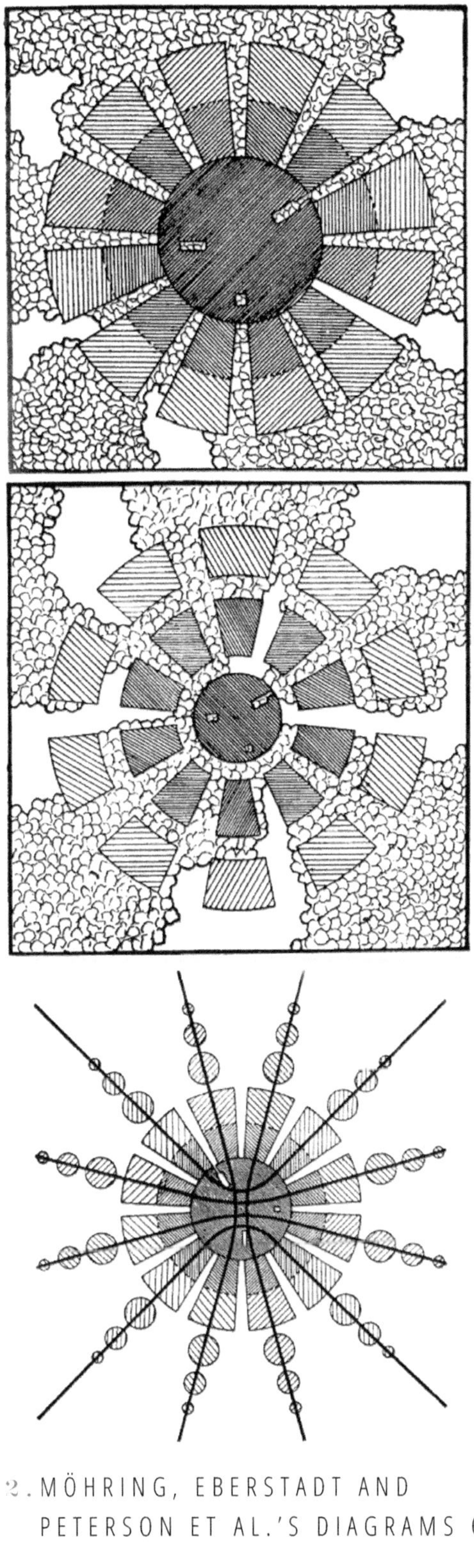

32. MÖHRING, EBERSTADT AND PETERSON ET AL.'S DIAGRAMS OF THE BERLIN METROPOLIS

with urban components of the environment. It also resulted in a definitive 'blue-green' plan for Singapore, along with streetscape greening plans. Attention also became focused upon the quality of landscape rather than simply its quantity. This change can be seen immediately, for instance, in the variety of plants in streetscapes and in park connectors conveying a continuous network of green spaces, in lieu of more fragmentary patches. The strategic thrusts behind a City in a Garden were six-fold. First, was to establish world-class gardens. Second was to rejuvenate urban parks and enliven the streetscape. Third was to optimize urban spaces for greenery and recreation. Fourth was to enrich biodiversity in an urban environment. Fifth was to enhance competencies of the landscape and horticultural industries. Finally. it was to engage and inspire communities to co-create a greener Singapore.[18]

A further turn away from Singapore as a Garden City and as a City in a Garden is the recent re-casting of the city-state as a 'City in Nature'.[19] In this scenario the vegetated aspects between and among buildings are seen to be more typically tropical and naturally emblematic of the primeval forests of the pre-colonial era. Concomitantly, road rights-of-way within urban areas are to shrink, making way for transit and for larger, somewhat more unruly landscapes to form. Simultaneously, the amount of wildlife, including animals, birds, and insects, are perceived as conforming to natural cycles of growth and development, as well as becoming more visible. An aim is to make Singaporeans more comfortable with day-to-day contact with nature and to strengthen the relationship between the natural and constructed aspects of the city's urban environment. Of course, the transformation also trades on those juxtapositions of 'gardens' and 'city' that went before. In this final rendition, if it comes to pass fully, all will be somehow present and the tropical character of Singapore will be more emphatically evident, as well as sensually and visually present.

33.VISION OF A CITY IN NATURE

C.

METAPHORICAL PERSPECTIVES

As a metaphorical perspective about habitation and, indeed, versions of civilizations, 'city' and 'garden' as broad categories have become intertwined, with the concepts of one term resonating against concepts of the other in order to somehow frame an appropriate and desirable path to be followed. A metaphor, after all, is when something is regarded as being representative or symbolic of something else, especially something abstract or complex. According to some scholars the word 'garden' is more an artifact of speech, whereas 'landscape' is associated with text. Nevertheless, the use of either or both terms can be the 'silent language of imperialism' as it tends to unify and naturalize the world to which it is applied. There, the tropes and narratives of garden and landscape tend to fall into several categories. The first are 'narrative tropes' involving conventionalized settings linked with particular events, which evolve repeatedly in a culture and that are also often associated with a nostalgic past of harmony with nature. In particular pastoral tropes are examples of this association. The second category involve 'genre narratives', usually associated with origins, foundational myths and foci on places.[20] Ideas, for instance, of Eden, Paradise and similar restorative narratives quickly come to mind. In many, if not most uses, of such metaphorical perspectives the aims are bound up with simplification of otherwise complex conditions and circumstances for the purposes of control or imbuing political meaning and for identity construction. Here, there is nothing necessarily nefarious about such deployments. As anthropologists often tell us we do tell ourselves stories in order to make ourselves feel alright. It is a very human manner of coping and of explanation.

One of the better known narratives of this kind is American pastoralism and the pastoral idyll. As one observer commented, "America was discovered before it was conceived...To become a reality America had to become a state of mind as much as a place, an entity whose identity and existence depended upon its meaning".[21] To the early settlers it appeared as a green gaze of virgin land, both magnificent and terrifying. This frightening 'New World' shared but also proved to be the grounds for reconstruction of a renewed civilization. Such vastness, however, lacked a meaningful context, making necessary the projection of a new yet familiar meaning. The religious motif of a 'New Jerusalem' allowed the early settlers to establish a mythical vision of America where the landscape would conjure up reassuring biblical images. The latter, in turn, would project the early settlers toward a future in which garden and city would be in harmony. The poetic descriptions of newly discovered America thus strengthened the idea of the 'New World' as a place where urban-pastoral society could live.

34. AMERICAN PASTORALISM: RICHMOND FROM THE HILL ABOVE THE WATERWORKS.

35. A MOTIF OF THE MODERN TECHNICAL TEMPERAMENT

The American urban-rural dichotomy is thus reflected in the mythical, ideal and metaphorical reality of 'Pastoralism' and its later dialectic with the so-called 'Modern Technical Temperament'.[22] Both of these fictitious and constructed visions of American splendor served the gradual construction of an identity of the American people, perennially divided, or so it seemed, between the idyll of boundless conquest of nature and the explosive force of technological advancement. In effect, Pastoralism proved to be an ideological motif that tended towards an idealized dimension of habitation that exceeded daily reality. Moreover, here the origins of Pastoralism at large date back to 'cultural primitivism' exemplifying the knowledge that derives from contact with nature but also with the moral and spiritual growth of the subject living in natural circumstances. Furthermore the concept of Pastoralism has at least two different connotations. First, it can be 'popular and sentimental' meaning that the primitive and rural realities of the countryside contrast sharply with those of the city. Or it can be 'imaginative and complex' by acting as an element of mediation between civilization and nature, the city and the countryside, nature and art, and so on.[23] In this form, according to some historians, Pastoralism represents a form of 'semi-primitivism'. In fact, it mediates more or less halfway between nature and civilization. It is, in short, "a third term between humanity and nature that is the identity between people and the natural world".[24]

With the advent of mass production, the United States moved towards a 'modern technical orientation' and temperament and the notion of Pastoralism became opposed to it as it linked to the specificity of place, production and so on, and away from simply the experience of civilization and nature. This gave rise to Modern Pastoralism which developed more palpably in the broad swaths of suburban metropolitan development during the mid-twentieth century. However, Modern Pastoralism, as such, cannot be considered a utopia but rather as an ideology. Essentially, it defined the symbolic functioning of the cultural artifacts of a 'Middle Landscape' in the form of a tradition of the past as a symbolic mask of the commodification that was undercutting its true essence.[25] The more egalitarian ideals of a heroic past, real or implied, also masked the regressive social realities of the time, when segregation and economic elitism were dividing the country. "Drawing on the promise of an open landscape and pastoral sensibilities and rhetoric, proponents of this perspective have linked city and countryside to describe or envision America as a healthy, harmonious and urban-pastoral society combing the best of both worlds".[26]

36. COMPLEX PASTORALISM IN THE LACKAWANNA VALLEY

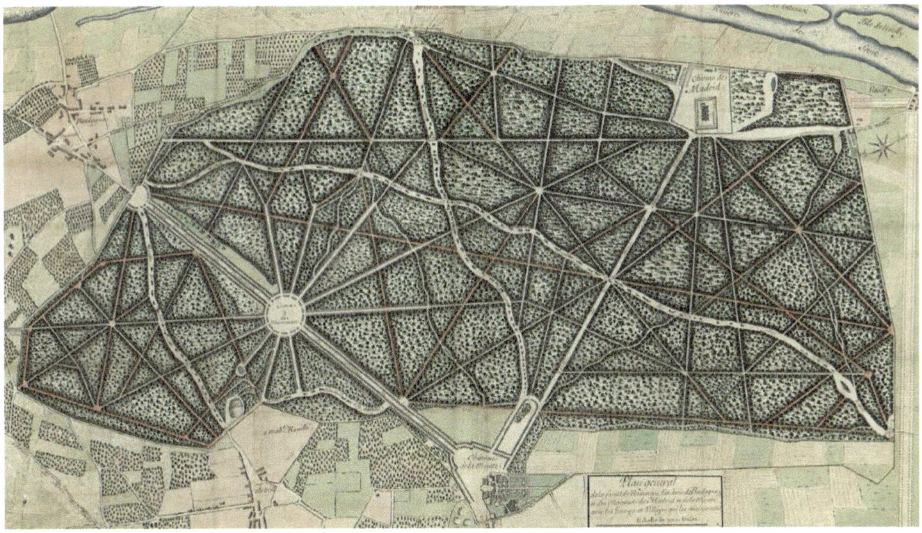

37. FOREST IN THE CITY – CITY IN THE FOREST

In Europe, earlier on during the eighteenth century, the city was already seen as a 'natural' entity and, more specifically, compared to a forest. Marc-Antoine Laugier, for instance, wrote in his essay about architecture of 1753 that "we must look at the city as a forest" and his 'primitive hut' also defined the universal and natural origins of architecture, placing itself as a universal foundation myth.[27] Indeed, this image of a city as a forest persisted, so that almost two centuries later in 1955 Ludwig Mies van der Rohe was to say "there are no cities, in fact, anymore. It goes on like a forest. That is the reason why we cannot have the old cities anymore, that is gone forever, planned city and so on."[28] The medieval city centre of Paris at the beginning of the nineteenth century probably appeared as a forest although more likely a jungle to Baron Haussmann, when he decided to regularize the urban pattern and to introduce tree-lined avenues. His plan for Paris introduced a new order to the city through greenery, a new theatrical quality that was composed by a sequence of urban scenes: the court, the garden and the forest. In fact compare Haussmann's city making to Le Nôtre's gardens of Versailles, where nature was seen as an almost solid mass through which pathways and fountain areas could be cut.[29] Furthermore, early designs for the Bois de Boulogne on the outskirts of Paris recall the urban interventions of Haussmann in these regards. Garden or park design, therefore, could be seen as a metaphor for city making as the city was treated as a forest or jungle that needed to be rationally and healthfully organized through a set of boulevards and avenues. The abundant use of trees inside the city also reinforced the idea of a city as a large garden and park. Moreover, this process of rationalization took place at different scales throughout nineteenth century Europe. Green spaces were a part of private villas at a small scale, large urban parks became a reality for all cities and pleasure gardens were provided hosting different urban social classes.

38. HAUSSMANN'S BOULEVARDS IN PARIS

In Italy, as Carnemello notes, "a characteristic of the garden and its space is to always represent a transitional element from various other aspects of the construction of the territory. Although reflecting urban culture, the garden is a transition between architecture and nature, between city and countryside, a transition between the house, the countryside and the wild nature – parts related to each other by gradual passages."[30] In fact, the Italian garden has historically united both *venustas* and *utilitas*, as vegetable gardens were often part and parcel of villa amenities since the ancient Romans. This connects back to the garden as a unifying element between different territorial scales. Again quoting Carnemello, "if we consider the garden as the highest formal expression of agriculture, it unequivocally belongs to the process of urban expansion, as both a complement to the urban palazzo in the case of the villa – expression of a singular form of expansion of the city in the countryside – and as part of the same process of formation and growth of the city, in the relationship between built spaces and free spaces, whether or not these are arranged as gardens." [31] Allegories about 'good government' and 'clean living' were also played out against images of the city - *citta* - as a vibrant organized whole and the countryside – *campagna* – as a verdant cornucopia of nourishment and sustenance. Ambrogio Lorenzetti's murals in Siena's Palazzo Pubblico are evidence to this effect.

Being rooted in the Italian tradition, the interaction and even unitary conception of the residential building and its garden explains why the Garden City Movement was embraced by neo-industrial and paternalistic authorities for the making of new neighborhoods on the outskirts of major Italian cities during the twentieth century. These garden neighborhoods – *quartiere giardino* or *borgate giardino* were considered successful as they were commensurate with national traditions of city building, rooted in principles of community and mediation between the town and the countryside.[32] The Quarterie Garbatella in Rome of 1920 to 1922, for instance, is an early example and is composed of dwellings distributed over two stories – *villini* – each with its own private garden and vegetable patch. The dwellings themselves were faithful to a picturesque vernacular vocabulary called baracchetto, or little baroque, which further underlined the overall nostalgic tone of the development. Similarly, the Citta Giardino, again in Rome of 1920 along the Aniene was an even bigger and more literal Garden City. Both examples and others like them were successful cases of the mediation between the city, the cultural values embedded into the Italian landscape and the tradition of garden design.[33]

39. THE ALLEGORY OF THE EFFECT OF GOOD GOVERNMENT ON CITY AND COUNTRY

40. GARBATELLA, ROME

d.

SINGAPORE'S CHANGING METAPHORIC LANDSCAPE

Following its early settlement during the first decades of the nineteenth century, Singapore went the way of many British colonial possessions with a focus on trade but also on local agricultural production. This was so much so that Singapore tragically lost almost all of its remarkable primeval forests and natural vegetation as noted earlier. In certain painterly images and cartographic representations of the day, an almost English countrified view of the broader landscape emerges with open fields, croplands. Patches of trees and homesteads dotting broader vistas moving well out beyond into the island's hinterland. This imagined rather than fully real perspective, for nothing could have been further from the diminishing reality of the tropical forest, exemplified a nostalgia for the home country far away and helped sustain a narrative trope about farming and living on the land. Not pastoralism like the Americans, for Singapore was more about crops, this state was, nonetheless, as if the landscape of the island was truly being remade. It was less the case of a conceptual binary between 'town' and 'country', as it was an almost complete makeover of the 'country' component.

3.1. SINGAPORE'S EARLY COUNTRIFIED LANDSCAPE

As mentioned about Singapore during the past half century or so, the coupling of terms qualifying city has moved from 'Garden City', implying putting the 'garden' component into the city; to a 'City in a Garden', which more literally describes the idea of a city within a garden; and then on to a 'City in Nature', where the notion of garden is re-qualified as wilder, more natural and presumably heading back towards the original primeval forests. At much the same time the concept of 'city' has been expanding in function, ambition and diversity. Moreover, as with other metaphors involving 'green' components, by naturalizing complex conditions, problems become seemingly simplified, the identity narrative of Singapore is reinforced, a promise of entertainment or recreation is provided, and economic profit is fostered. Furthermore, the trend from Garden City to a City in Nature mirrors the circumstances of Singapore during this time. On the heels of vast destruction of natural and garden-like circumstances, as well as rounds of cleaning up, bringing the garden into the city would seem to follow both logically and practically. With tree-planting and the like, it would also engage many Singaporeans in a joint effort and begin to convey a sense of pride in their environment. Furthermore, as the Housing and Development Board's program gained momentum and success, the sheer need to offset the built and constructed environment with something more natural, and no doubt softer, became more and more obvious. Then because of the persistence of these efforts, Singapore became tangibly greener to the point where the scale tipped, as it were, to Singapore being able to be perceived as a garden more broadly into which city building was occurring. The most recent turn to a 'City in Nature', although not fully realized, points to the special character of Singapore as a place, within the branding that is also taking place, and the authenticity and distinction which needs to be more fully involved. Singapore is, after all, tropical and with the idea of genre narratives, introduced earlier, this implies a certain kind of nature. It is one that is wilder, more jungle-like, biodiverse, and sensually engaging.[34]

In the story and intertwining of 'blue' and 'green' the idea of nature also resonates better with today's circumstances and times, strongly tinged politically and otherwise with the goodness and righteousness of sustainability, ecological relationships and even senses of survival. Also, while on the matter of water independence and sustainable sources, the hydrologic and water re-use cycles engage a certain autonomy, natural circumstances, biodiversity and how the 'green' relates technically to the 'blue', as it were. Further, it can result in less costly maintenance and enhancement of the environmental 'friendliness' of the operations of the water systems.[35] Certainly, many individual families cannot exert the same impact as a more pervasive natural phenomenon. Nor can an expansive garden survive without the strong ecological underpinnings to be found in nature. In this trend, however, several consequences and issues can be seen to occur. First, the acceptance of a 'City in Nature', or a literal state close to it by Singaporeans needs to be cultivated and developed. Anecdotally, it seems that many enjoy attractions and benefits of nature but also dislike and wish to avoid less attractive aspects. As one naturalist observed, while people like butterflies and butterfly gardens, they do not like the caterpillars that accompany them.[36] Second, particular forms of buildings and public infrastructure are clearly going to be favored over others in making a "City in Nature', including green wall and roof structures, dedicated traffic lanes for cyclists, pedestrians, joggers and people out in nature generally. Also the width of some roadways will likely shrink and public open spaces will be rendered in more natural ways. With the right kind of adaption by inhabitants, all of this could be and probably will be an exciting, healthful and trend-setting condition. Given the constant tropical circumstances of Singapore, it may also mean that the isolation of communities may lessen or become more consolidated with concomitant shifts in fruitful use patterns and arrangements of live, work and recreate.

The metaphor, if not reality, of a 'City in Nature' also presents a segue into the concept of 'hyperobjects' and those entities of such vast temporal and spatial dimensions that they defeat traditional ideas about what a thing is in the first place. Immediate examples include the health of the biosphere and events like climate change. The term is used by object-oriented philosophy adherents like Timothy Morton.[37] It was first coined in 1967 in computer science to refer to n-dimensional, non-local objects. In the work of Morton the concept is brought to bear to problematize environmental theory from the standpoint of ecological entanglements and overcoming the bifurcation of nature and civilization, or the idea that nature exists as something apart from but sustaining society. By contrast, the claim is made that we are embedded in nature. Morton also uses the term 'dark ecology' to apply to the irony, ugliness, and horror of ecology, as well as the term 'nets' to refer to the interconnected uses of all living and non-living things including the idea of infinite connectedness and infinite distinguishing differences.[38] Taken in *extremus*, Singapore's 'City in Nature' could be seen to participate in a hyperobjective state, especially with the emphasis on ecological circumstances, natural progressions of things and relationships with and among people and things. Moreover, 'clean and green' seems likely to take on a new meaning in the sense of being unadulterated and completely natural, including potentially elements of Morton's 'dark ecology'. The early specter of this eventuality would also make Singapore unique in the world. It would remain to be seen, however, whether such a state would have the desired edifying effects on Singaporeans as the earlier metaphors of combinations of 'city', 'garden', and 'green'. Many probably think not. On par though, in the span of a little less than two centuries Singapore has moved, metaphorically speaking, from attempts at an almost total makeover of the island in the image of an English countryside on to the quintessentially modern idea of the 'Garden City' and now on to the less conceptually bifurcated and more complicated relational complexes of a 'City in Nature', with its biophilic and potentially hyperobjective associations.

42. THE UNCANNY INTIMACY AND STRANGENESS OF HYPEROBJECTS

endnotes

1. Lee Kuan Yew. 2012. *From Third World to First: The Singapore Story 1965=2000, Memoirs of Lee Kuan Yew, vol, 2* (Singapore: Marshall Cavendish), p.200.

2. Boon Siong, Neo. 2009. *A City in a Garden – Developing Gardens Singapore: Centre for Governance and Leadership,* (Civil Service College), p.2.

3. Lee, p.309.

4. Lee Kuan Yew. 1968. Opening Speech of the "Keep Singapore Clean" (Singapore Conference Hall, 1st October, National Archives of Singapore).

5. Boon Siong, p.5.

6. Boon Siong, p.6.

7. Lee, 2012, p.300.

8. Lee Kuan Yew, 2011. *Hard Truths to Keep Singapore Going.*

9. Boon Siong, p.78.

10. Chou, Cynthia. 2014. "Agriculture and the End of Farming in Singapore" in Timothy P. Barnard, ed. *Nature Contained: Environmental Histories of Singapore* (Singapore: NUS Press), p.233.

11. Chou, p.234.

12. Howard, Ebenezer. 1965 (1890). *Garden Cities of Tomorrow* (Cambridge, Mass: The MIT Press).

13. Miller, Melvin. 2010. *English Garden Cities: An Introduction* (Chicago: University of Chicago Press).

14. Richie, Alexandra. 1998. *Faust's Metropolis: A History of Berlin* (New York: Carroll and Graf Publishers, Inc).

15. Sutcliffe, Anthony. Ed. 1984. *Metropolis 1890-1940* (Chicago: University of Chicago Press).

16. Neo, B.S., Gwee, J. and Mak, C., "Growing a city in a garden", in Gwee, J. (Ed.), *Case Studies in Public Governance - Building Institutions in Singapore* (Singapore: Routledge, 2012).

17. Boon Siong, 2010.

18. Interview with Dr Lena Chan and Mr. Lim Liang Jim on 22 June 2017

19. Yeo, H. and M Chng. 2017. *Building a City in Nature* (Singapore: Centre for Livable Cities, CLC Insights 24).

20. Potteiger, M. and J. Purinton, 1998. *Landscape Narratives, Design Practices and Telling Stories* (New York: John Wiley and Sons, Inc).

21. Machor, J. L. 1987. *Pastoral Cities, Urban Ideals and the Symbolic Landscape of America* (Madison: The University of Wisconsin Press), p.27.

22. Rowe, Peter G. 1991. *Making a Middle Landscape* (Cambridge, Mass: The MIT Press).

23. Marx, Leo, 1964. *The Machine in the Garden: Technology and the Pastoral Ideal in America* (New York: Oxford University Press).

24. Rowe, p.221.

25. Rowe, p.200.

26. Machor, p.5.

27. Laugier, Marc-Antoine. 1753 (1977). *An Essay on Architecture, Translated by Wolfgang and Anni Hermann* (Los Angeles: Hennessey and Ingall).

28. Interview with Mies van der Rohe by J. Peter in Lambert, Phyllis. ed. 2001. *Mies in America* (New York: Henry N. Abrams).

29. Jordan, D. P. 1995, *Transforming Paris: The Life and Labors of Baron Haussmann* (New York: Free Press).

30. Carnemolla, Adriana. 1989. *Il Giardino analogo, considerazioni sull'architettura dei Giardini* (Rome: Officina Edizioni), p.42.

31. Carnemello, p.180.

32. Tagliaventi, G. 1995. *Citta Giardino: Cento Anni di Teorie Modelli Esperieze* (Rome: Gangemi Editore).

33. Tagliaventi, p.30f.

34. Potteiger, M. and J. Purinton, 1998 and Yeo and Ching, p.5f.

35. Interview with Professor Leo Tan and Dr. Darren Yeo, Biological Sciences Department, NUS on 17 August 2017.

36. Interview with Mr. Khew Sin Khoon, CEO of CPG Consultants on 8 February 2017.

37. Morton, Timothy. 2013. *Hyperobjects: Philosophy and Ecology after the End of the World* (Minneapolis: University of Minnesota Press).

38. Morton, p.65f.

chapter

04

WATER RESOURCES AND SUSTAINABILITY

P. G. Rowe and L. Hee, *A City in Blue and Green*,
https://doi.org/10.1007/978-981-13-9597-0_4

In addition to 'green and clean' in the service of national identity, equality and attractiveness, sustainability of Singapore with regard to water supply and use is of strategic national concern. Indeed, it is expected to achieve this by 2061 when Singapore's remaining water agreement with neighboring Malaysia will end, even though imported sources from Malaysia account for 40 percent or so of overall supply as of 2009.[1] Moreover, sustainability in this regard is so strategically important that all other matters of national concern 'bend at the knee', so to speak, in abeyance to this single outcome and issue. At present Singapore is dependent upon what are called its 'four national taps' as its supply strategy. In addition to importation, 'local water catchment' provides some supply, largely by way of collecting and channeling rainwater to Singapore's 17 reservoirs via a network of rivers, canals and drains. Although water rich in terms of annual rainfall, unlike other places in the world, like Boston in the U.S. or Melbourne in Australia, with their abundant resources in dams in outlying areas, also at higher elevations than the cities proper, Singapore has a limited land and catchment area. Therefore, more conventional water storage, treatment and supply approaches are not feasible in face of the city-state's demand. Two compensatory sources of supply are so-called NEWater, Singapore's brand of reclaimed water, involving membrane and other cleaning technologies, and desalination of seawater. On the demand side, the government also actively promotes improved performance by way of active and successful demand reduction, largely in the form of individual conservation at the source of water use. This could be imagined as the fifth National Tap, an equally critical component to achieve sustainable water supply in the long run. Indeed, today Singapore is getting closer to the goal of 130 liters per person in daily use by 2030. The real genius and innovation, however, in Singapore's domestic water supply approach comes not so much from specific technical innovation as by way of overall conceptualization.

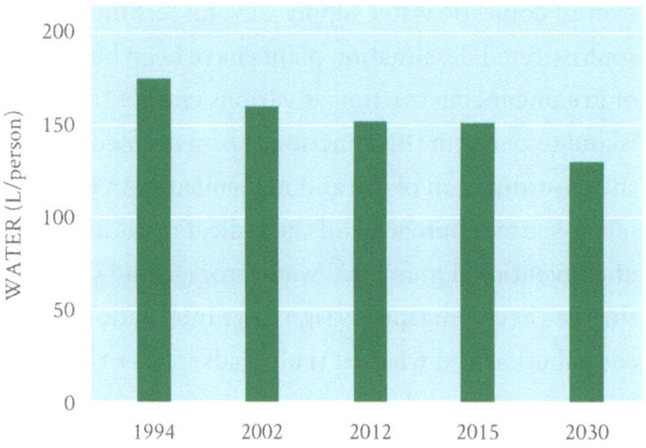

43. SINGAPORE'S DOMESTIC HOUSEHOLD WATER CONSUMPTION

Instead of a large raw storage, treatment and supply system as elsewhere, use is made of a closed loop in conjunction with 'used water' instead of 'waste water' that is constantly re-cycling.

The colonial legacy of a separated storm water and sewage system of conveyance effectively sets the foundation for Singapore to continue to expand the drainage and sewerage network across the island. The fundamental principles of the closed loop system are to gather almost every drop of rainwater runoff, collect every drop of used water and recycle every drop of water more than once, thus obviating the need for large storage devices which Singapore lacks as a land-scarce city-state island. In terms of the energy-water nexus, the sewage stream is being re-cycled in the form of carbon extraction from waste and co-generation of electrical power to augment current capacities and help drive the requisite pumping required in the re-cycling efforts. This augmentation, of course, also helps lessen the load on imported sources of energy, like oil and gas, where Singapore continues to be dependent. Though not completely self-sufficient, the trading of one dependency in water for another in energy resources is clearly lessened. Finally, desalination and together with NEWater production further augments and 'tops up' so to speak, necessary overall domestic water supply. Several technically sophisticated desalination plants have been built or are under construction at various ends of the island to perform this function. As mentioned it is the co-ordination of use and implementation of this multi-source approach and the radical re-thinking of conventional municipal water storage and supply that has been Singapore's signature international contribution and where it truly leads, rather than in the membrane, desalination and other treatment technologies *per se*. Furthermore, although not completely implemented at this writing, sufficient proof of concept and several working plants and conveyance systems will ensure success, probably well before the 2061 deadline. In addition, significant aspects of the radical overall approach seem to be transferable to other water-scarce urban situations elsewhere in the world. What is excluded from this depiction, however, are virtual water flows ascribed to Singapore from elsewhere in the world in order to support agriculture and industrial products used by Singaporeans. While likely to be sustainable with regard to important elements of domestic demand like household use, the city state is not and likely will not be fully sustainable with regard to virtual water consumption, like many other nations in the world.

a. SINGAPORE'S CONSTRAINTS AND OPPORTUNITIES IN WATER RESOURCES

Largely due to its relatively small size and lack of a full complement of natural resources, Singapore faces a number of constraints as it attempts to achieve water sustainability by 2061. By contrast, there are also some advantages or opportunities that can be exploited further. At present the land area of Singapore is 722.5 square kilometers, including some 140 square kilometers of reclaimed land, or around 24 percent of the land area since 1959, constructed mainly in the west and east.[2] Despite this relatively small area and dense land occupation at around 7,042 inhabitants per square kilometer, Singapore has its rainwater and run-off storage capacity appreciably through construction of reservoirs in the centre of the island and at river mouths. Early stages of this were described in chapter 2, but more recent advances have been made from fourteen reservoirs in 2005 to seventeen reservoirs as of today.[3] With the recent building of the Marina Barrage and both the Punggol and Serangoon Reservoirs the drainage catchment area now covers roughly two-thirds of the national territory.[4] By contrast, Melbourne in Australia, another city with a comparative emphasis on 'green and blue' developments leveraged its spatial advantage to easily tap into its protected and forested catchments to harvest water stretching along the Great Divide, a substantial area in the middle of the State of Victoria. Recently, however, climate change and drought conditions revealed some vulnerabilities with regard to adequate supply, despite the large latent capacity of the catchment areas.[5]

Unlike other large cities, Singapore does not enjoy abundant groundwater supplies. Indeed, as described earlier, serious consideration of groundwater as a potential source of constant supply has been ruled out, even though the PUB continues to monitor and look into the possibility.[6] By contrast annual rainfall is relatively abundant at around 2,400 mm per year in Singapore's tropical location, as is sea water, surrounding the island. One issue, however, with the tropical monsoonal conditions are the peaks and valleys in precipitation, where supplies can dry up between bouts of intense rainfall and catchment. Longer term droughts, like the one that afflicted the island in 1963 may be uncommon but cannot be ruled out entirely. The weather can also influence the relative salinity and other chemical compositions of reservoir and estuarine water services as well, posing potential problems for reliable treatment processes. As a corollary to this, fluctuations in electrical demand associated with desalination processes and at varying levels of salinity can effect operating efficiencies of treatment plants. Clearly Singapore must make a trade-off between largely imported and often costly energy resources in its water sector, especially with the use of desalination technologies. By contrast, though, the city state benefits from the British legacy of a separated and dual storm-water and sewage system, particularly with regard to down-line treatment. At the moment energy requirements per cubic meter for domestic water treatment is about 0.66 kwh per cubic meter of water, but is expected to double with technical

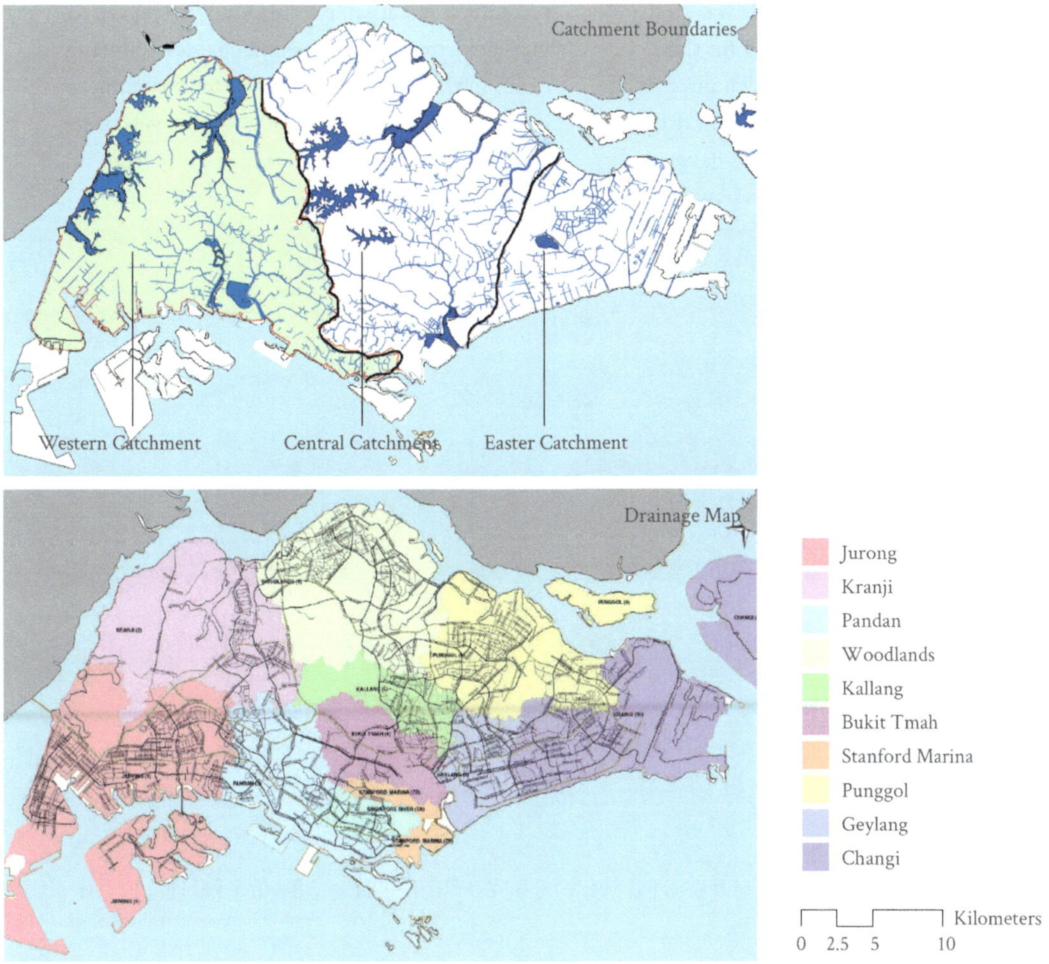

44. MAPS PERTAINING TO SINGAPORE'S CATCHMENT AND DRAINAGE SYSTEM

shifts towards desalination and water reuse, before reaching a level of something like 0.75 kwh per cubic meter in the longer term.[7] To put this in context, it has been estimated that about two percent of the Singaporean electricity demand is already dedicated to water and treatment processes.[8]

Singapore's concerted efforts towards sustainability in domestic water consumption, however, are dwarfed by the imputed virtual water components of food and industrially produced products. At present this seems to amount to about 20 times the current blue water supply. Here virtual water is a term that embraces the amount of water used to produce a certain good or product. It was first introduced during the 1990s to account for the limitations of water scarce regions which imported food grain grown with other's water.[9] From this the broader concept of a water footprint was developed, allowing calculation of the amount of virtual water in commodities of various kinds, such as the ratio of the total volume of water used to the quantity of production.[10] More basically, a water footprint has three components. They are: the green water component or the volume of rainwater evaporated or incorporated into products; the blue water component or volume of surface or groundwater evaporated or incorporated into a product or returned to other catchments or the sea; and the gray water component, or the volume of water needed to dilute loads to such an extent that they reach certain water quality standards.[11] Anyway, Singapore's small scale, particularly in relationship to its largish population effectively precludes it reaching a sustainable condition with regard to virtual water. As described in chapter 2, during times in the past Singapore was far less numerous in population and did, in fact, reach levels of self-sufficiency in food production, mainly from small family-owned farms.

	2010	2060
DOMESTIC	45%	30%
NON-DOMESTIC (POTABLE)	33%	42%
NON-DOMESTIC (NON-POT.)	22%	28%
NON-DOMESTIC (TOTAL)	55%	70%
PER CAPITA	154 LITRE/DAY	130 LITRE/DAY (2030)
TOTAL (MILLION GALLONS/DAY)	380 (100%)	860 (100%)

45. PROPORTIONS OF TOTAL AND SEGMENTED WATER USE, 2060 PROJECTION

b.

SINGAPORE'S FOUR NATIONAL TAPS

On the resource side, Singapore relies for progress towards water sustainability through its four taps, plus concerted efforts to manage local water consumption.[12] The first tap is supplied by importing water from Johor in Malaysia under a series of water agreements, briefly described earlier. This tap is estimated to provide 40% of Singapore's water demand, corresponding to 250 million cubic meters per year as assumed by Lenouvel et al. in 2013.[13] In this process quality drinking water is produced from freshwater by way of conventional water treatment processes. Despite having the source of imported water, the Singapore government felt the need to be more sustainable, and had begun to intensify its efforts in the 1970s to diversify the sources of water supply. This is being undertaken through the establishment of the additional National Taps.

The second water tap relies on rainwater catchment and Singapore's rainwater and run-off storage system described earlier. A clear aspect of the viability of this tap also goes back to the clean-up programs of the early Republic, described in chapter 2 and to Lee Kuan Yew's abiding interest in Singapore being 'clean and green'. Today development of harbor and industrial activities are concentrated in the south and southwest of the island and away from the main water courses and reservoirs. Singapore has successfully developed a total of 17 reservoirs and expanded the water drainage catchment area to two-third of the nation's land area. The energy footprint of this tap is also quite small at around 0.25 kwh per cubic meter of water.[14]

The third water tap has relied, since 2003, on water reuse and the so-called NEWater Initiative. This currently can supply up to 40 percent of Singapore's water needs. The NEWater journey started way back in the 1974 when then-Public Utilities Board started experimenting on water reclamation and treatment, however, it did not take off because the technologies then were too costly and unreliable. It was only in the 1990s when technological advancements had brought down the cost of membranes, that PUB revisited the possibility of large-scale water recycling. Through overseas study of other cities' technologies and thorough feasibility study, the engineers, in 2000, eventually designed and constructed a demonstration water reclamation plant in Bedok, a milestone that led to the creation of NEWater, Singapore's brand of reclaimed water. Fundamentally, NEWater involves a production process for treating used water, including effluent, through a succession of stages, resulting in drinkable water meeting the highest of standards. The first stage of the process is known as microfiltration.[15] It is when the treated used water is passed through membranes to filter out solids, colloidal particles, disease-carrying bacteria, some viruses and *protozom cyots*. The filtered water that goes through the membrane contains only dissolved salts and organic molecules. The second stage of this NEWater production process is known as reverse osmosis, where a semi-permeable membrane is used. This semi-permeable membrane has very small pores in it which only allow very small molecules like water molecules to pass through. Consequently, undesirable contaminants such as bacteria, viruses, heavy metals, nitrate, chlorides, sulfates, disinfection by-products, aromatic hydrocarbons, pesticides, etc., cannot pass though the membrane. In other words,

NEWater is reverse osmosis water and is relatively pure and contains negligible amounts of salts and organic matter. At the onset of the third stage, the water is already of a high quality and it acts as a further supply back-up to the reverse osmosis. In this stage ultraviolet disinfection is used to ensure that all organisms are inactivated and the purity of the water product can be guaranteed. Before storage of NEWater in tanks, the water balance of pH and the addition of some alkaline chemicals takes place to restore the acid alkali pH balance. It is then ready to be piped off to a wide range of applications mainly for non-potable use by the industrial sector such as wafer fabrication plants which require ultra-clean water. About 2 percent of NEWater is injected into reservoirs to blend with the raw water, before it is sent to the conventional waterworks for treatment and supply to customer taps.[16] Such a process of indirect consumption of NEWater, in addition to extensive public education and outreach efforts, is part of PUB's strategy to overcome aversion by the public to the reclaimed water.[17] The accompanying diagram illustrates the process of NEWater production. The tertiary wastewater treatment involved with NEWater consumes relatively high amounts of energy due to the intensification of the nutrient removal process.[18] In some places, for instance, such advanced wastewater treatment processes are highly energy intensive, with demand ranging from 0.39 up to 3.74 kwh per cubic meter of water. In Singapore the energy demand is on the order of 0.95 kwh per cubic meter of water, not very high but, in aggregate, significant.[19]

The fourth water tap relies on seawater desalination with, in Singapore, the commissioning of its first two plants in 2005 and 2013, with water production capacities of 50 and 115 million cubic meters per year.[20] Currently, the share of the desalination tap in water supply has been estimated to be approximately 8% of the total, with an assumption of imported water providing 40% of the water supply and rainwater harvesting at 30% .[21] The stated capacity of the three desalination plants combined amounts to 130 mgd that can meet up to 30% of Singapore's water demand,[22] but with high amounts of energy input at about 4.10 kwh per cubic meter of water, or about 74 percent of the total amount of energy expended by Singapore's water sector.[23] There are also different configurations of desalination. In the Singaporean context one of the most promising is the variable salinity desalination concept, devised to respond to the weather-variable conditions within catchment areas described earlier, especially in estuarine circumstances. There levels of salinity can vary substantially, such that harvesting needs to respond to the variation and in a timely fashion. Instead of remaining idle during periods of high overland flow and freshwater infiltration into a catchment area, the desalination mode is altered in response.[24] Overall, the process still involves screening, microfiltration, reverse osmosis and disinfection, as shown in the accompanying diagram, but with modes that respond to the levels of total disposable solids and salinity levels to product potable water that meets WHO and USEPA' Drinking Water Guidelines and Standards'.[25] The advantages of the variable approach include: much lower energy consumption through the mode switching and also significantly lower capital an operating costs.

Other versions of desalination technology consist of essentially two kinds. They are: reverse osmosis and combined electrodialysis-electrodeionization processes. Reverse osmosis has been around since the 1970s and was the technology, discussed in chapter 3, though defective, that gave impetus to Singapore's conviction to pursue non-conventional sources of water supply. The process involves forcing seawater against a water permeable membrane which prevents dissolved salts and minerals from getting through.[26] However, it has a couple of inherent economic and technical limitations that will curtail its future use. First, the feed pressures are very high requiring substantial energy. On average 3.5 kwh per cubic meter of water are required. Second, pressures

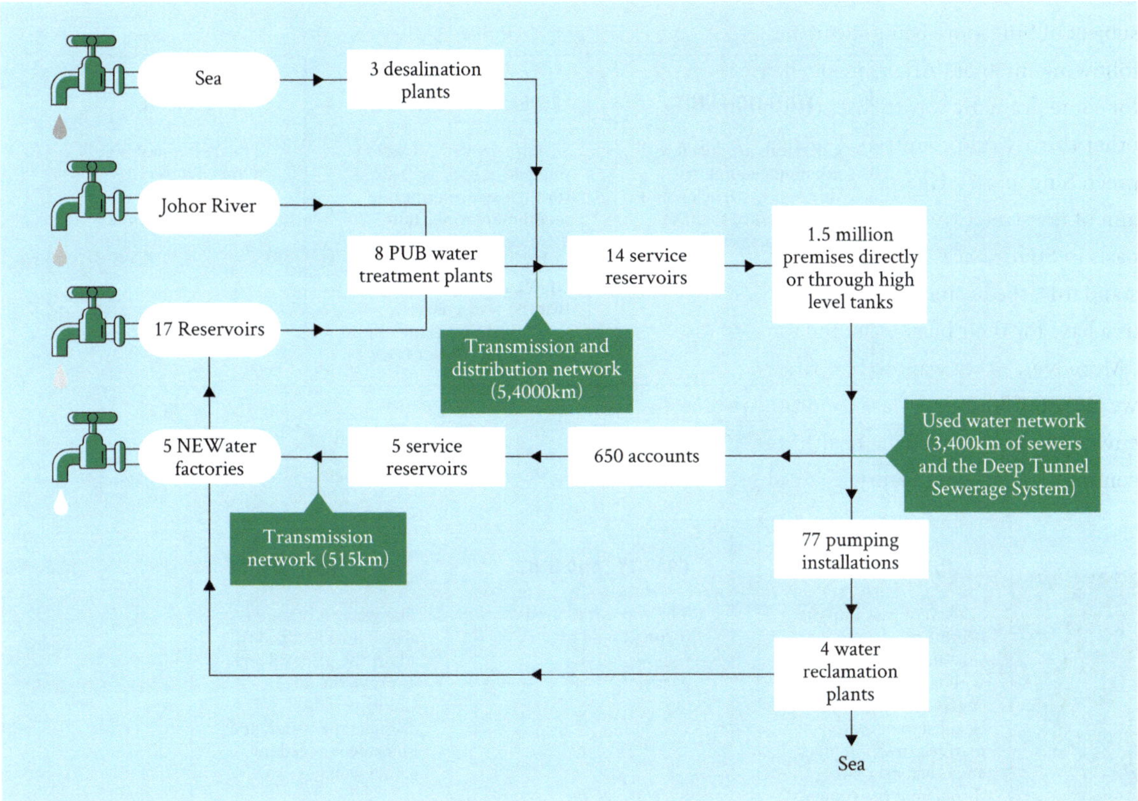

46. SINGAPORE'S FOUR NATIONAL TAPS

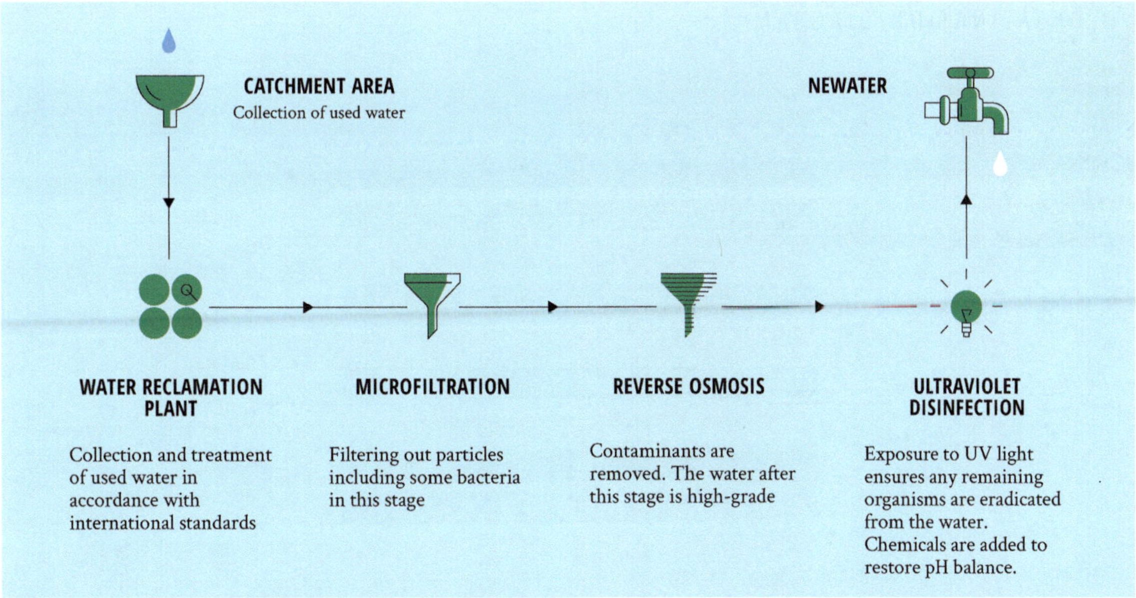

47. NEWATER DIAGRAM

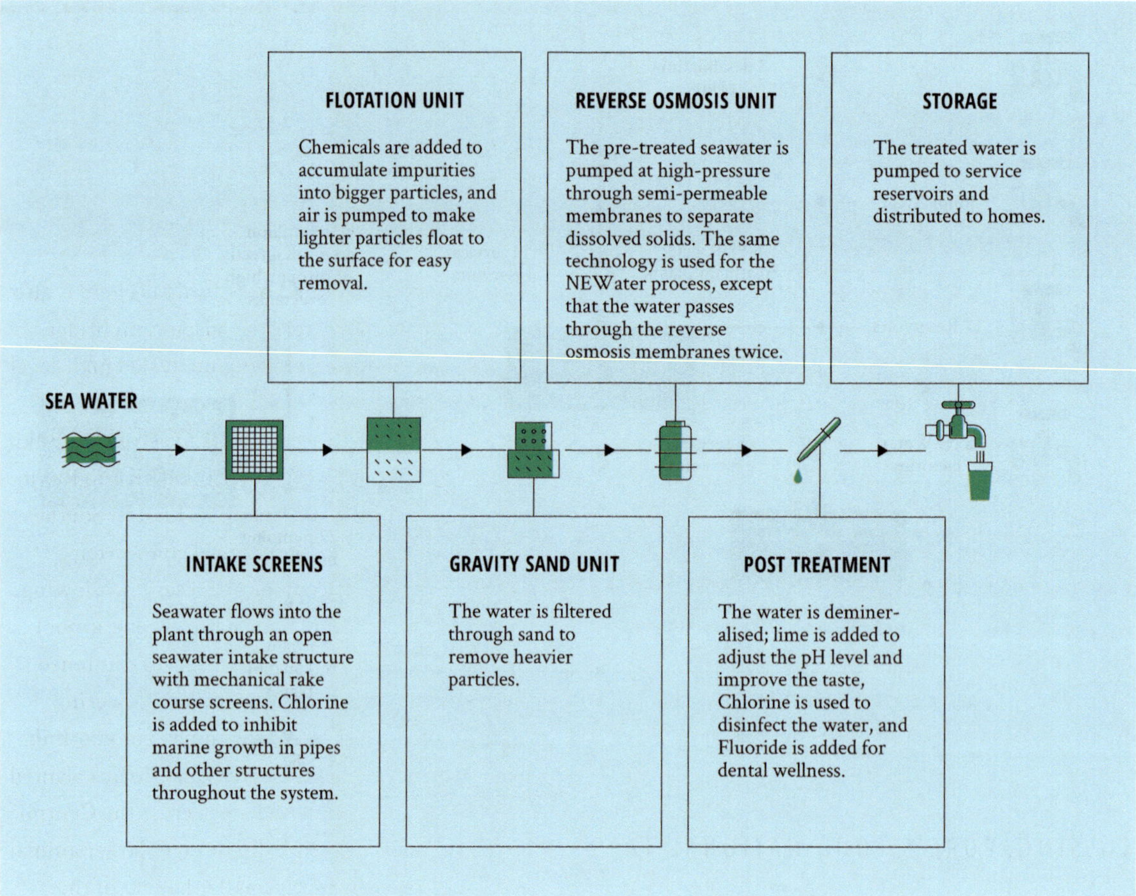

48. DESALINATION DIAGRAM

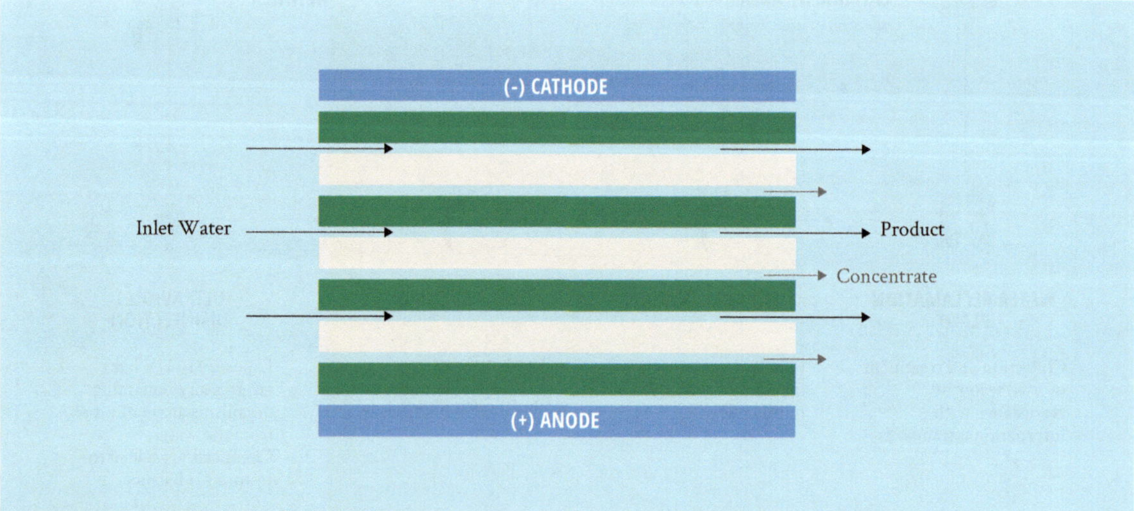

49. MICROFILTRATION SCHEMES AT WORK

can also cause serious deterioration of membranes requiring relatively constant replacement. By contrast the electrodialysis-electrodeionization schemes are far less energy consuming, testing at 1.8 kwh per cubic meter of water. In Singapore this technique was pioneered by Evoqua Water Technologies, previously known as Siemens Water Technologies, as a part of the 2007 'Singapore Challenge' competition which they established.[27] Technically, electrodialysis is an electrical separation process which selectively removes salt ions based on their electrical charge by transferring them through semi-permeable ion exchange membranes charges with a direct current voltage. The seawater separates into the product water with salt ions removed along with a concentrate with salt ions. However, flow through the series of electodialysis modules results in salt content that remains present but is too low for further removal by electrodialysis. Then electrodeionization takes up involving the transfer of solutions through semi-permeable membranes charged with electrical potential but with the use of exchange resins which increase the efficiency of transfer. The advantages of the combined electrodialysis-electrodeionization method over reverse osmosis are fairly obvious. It includes: lower energy expenditure, a tolerance of feed water of lower quality and a water product recovery rate that is higher. Further innovation of membrane technologies are also under serious investigation, including ceramic membranes and nano-filtration membranes, both of which seem likely to be cheaper to produce.

One of the main proponents and entrepreneurs in the water industry is Olivia Lum. She started off her business life as a child, working in a rattan factory and selling in the streets at an early age.[28] Determined to go into business she then trained as a chemist at the National University of Singapore, thinking this was a strong way of making a career in business. During the onset of the Asian economic crisis she worked as a chemist for Glaxo Pharmaceuticals and became interested in water technologies. In 1989 she left Glaxo and founded Hydrochem an early version of Hyflux, her current firm, using personal assets in order to raise the start-up capital for an enterprise in the Tampines Industrial Park. Starting off with only a staff of one clerk and one technician, she worked towards realizing the potential of membrane technology. In 2002, Hyflux was awarded a contract to build Singapore's third NEWater plant by the PUB in Seletar and in the following year, was awarded to build Singapore's first large-scale seawater desalination plant - the Singspring Desalination Plant - in Tuas. Subsequently, Hyflux again won the bid for the PUB's tender to build the country's second desalination plant. After some 29 years in business, Hyflux has successfully constructed some 1,000 plants of various sizes all over the world and helped put Singapore at the very forefront of the water treatment industry. Lum's efforts, though not literally inventing membrane technology and associated desalination technologies did put the general approach on the map, as it were, as far as large-scale projects were concerned. At present she is involved in producing her own brand of oxygen-rich water under the ELO brand and still concentrates on what for her are the two important aspects of the water resource business: cleaning and public health. One of the major hurdles that needed to be overcome was to gain the public's confidence in the safety and reliability of reused water.

In addition to these four taps, as mentioned, Singapore also embarked upon a strong program of water conservation and management, with the aim of reducing domestic household demand to an average of 130 liters per day in 2030 down from its current level of 151 liters per day per person (as of 2015).[29] Primarily conservation efforts have revolved around intense public education efforts, including exhibitions with elaborate explanations about the impact of various water-conserving technologies and strategies for individual and family use. Indeed, the PUB has been very active in these regards, especially among and with children a route that appears to catalyze further involvement among adults as well.[30] In the future, the present conservation goals appear to be realistic. In addition, among the taps, NEWater seems likely to account for 55 percent of domestic water needs by 2061 with 40 percent through five plants at the end of the year and desalination will likely supply 30 percent of demand by 2061. This will also add to the needed energy budget, though with greater efficiency. The Singapore Government aims to increase the water catchment area to 90% of Singapore's land area, up from about 66% of current water catchment area. Overall, the goal of sustainability by 2061, if not before, appears likely if not almost certain to be achieved.

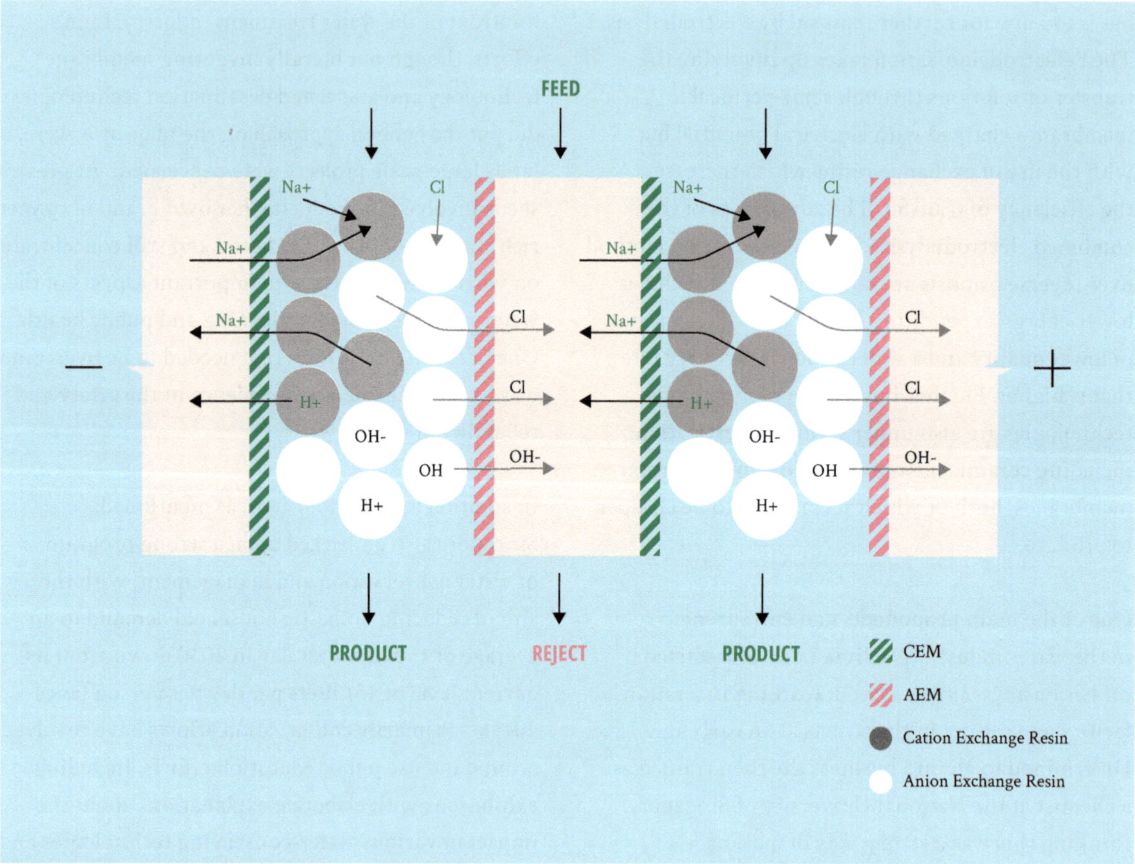

50. DIAGRAMS OF ELECTRODIALYSIS – ELECTRODEIONIZATION

51. TUAS WATER TREATMENT MEMBRANE TECHNOLOGY AT WORK

C.

SINGAPORE'S CLOSED LOOP SYSTEM

As indicated in the accompanying diagram, the crux and radical aspect of Singapore's approach to domestic water sustainability is the arrangement of the four tap technologies, plants and distribution networks into a closed loop system, where the Unaccounted for Water (UFW), mainly consists of water lost through leaks, stands at 5 percent, among the lowest globally. Consequently, the break with other conventional approaches to water supply is the constant re-cycling and reuse of water in the overall system, which theoretically remove the need for large areas of reservoirs and conventional treatment plants often adopted by many countries in the world. Also, as the network increases to serve expanded populations, so does the useable water storage capacity, making the entire system viable over time, in theory if not in practice. Central to Singapore's approach is naming and popular acceptance of 'used water', *in lieu* of 'waste water' and because of this, a capacity to re-cycle and reuse almost every drop of water on the island. Also key here is the harvesting and use of water in and from urban rather than more pristine circumstances. This is considerably different from conventional water supply regimes in places like Boston and Melbourne, where water harvesting is largely if not totally confined to catchment areas which are preserved as such and separated away from potentially contaminating other uses. This change in the spatial logic of water harvesting and potable water reuse is both a liberating and persistent feature of the Singapore approach. It does not mean, however, that the nation's catchment areas that have been well-defined and at least partially protected for years will fall away as sources of water supply for they will not. It also means that reservoirs for long-term diluting retention of large volumes of water, like the Marina Barrage, close to very urban adjacent land uses and sources for NEWater production will be maintained and become something of a new norm in the closed loop system of reuse and delivery.[31] In short, the indigenous three taps of Singapore's supply resources will remain with the emphasis in the use of one over the others determined by other related factors such as energy use and technological reliabilities.

The reduction of energy dependence in the water sector has been taken up by local authorities, largely under the umbrella of the Water Industry Program Office in partnership with the private sector and university research capacities. This involves at least three initiatives. The first is reduction of desalination energy needs through grant research by Evoqua Water Technologies among others. The second involves an increase of water catchment area to 90 percent of the island territory and with the use of variable salinity technologies to collect and treat water from estuaries.[32] The third is reduction of sludge processes, also involving pre-treatment from reverse osmosis by membrane bio-reactors. The PUB has also increased the integration of energy systems in water facilities again in three ways. First by involving increased optimization of biogas production from sludge digestion in used water treatment plants. Second by using production of energy through turbine technology in desalination plants. Third by offsetting operational costs of producing clean water and electricity through system integration between a power generation plant and a water reclamation plant. Independent study has shown that the long-term target of 0.75 kwh per cubic meter of water seems likely to be achieved, especially with suspension of reverse osmosis and reliance on breakthrough innovation in ion exchange and biometrics.[33]

As mentioned here, the evolution of the Singapore water convergence and treatment system has been incremental from its early trials and beginnings even as far back as the 1970s. Indeed, it is through the various steps in the process, the discoveries made and the successive improvements incorporated that Singapore has made an indelible mark on the contemporary enterprise of the urban water sector and water industry. Currently another instance of this is being made through the island-wide Deep Tunnel Sewage System (DTSS) under the auspices of the PUB. Before its design and implementation the used water system of Singapore involved six water reclamation plants and some 130 sewage pumping stations. With the DTSS the scattered pumping stations can be progressively phased out and replaced with a single large diameter tunnel which conveys used water flow by gravity. The used water is then pumped up only once at a centralized pumping station located at two water reclamation plants located at the eastern and western part of Singapore. The scale economy of this approach is cost effective compared to the older and existing infrastructure. By reduction of surface facilities it also frees up valuable land for other higher-valued development. It also ensures adequate long-term convergence and treatment capacity, while presenting a more robust, reliable and resilient used water system.[34] In effect, the DTSS is becoming the super-highway for used water. Construction of the DTSS is being pursued in two phases. The first phase, completed in 2008, comprised the Changi plant of 176 million gallons per day in the east of the island and is made up of 48 kilometers of deep tunnel sewers and 60 kilometers of linked sewers. The second phase, commenced in 2014, will feature a similar 176 million gallon per day plant plus 25 million gallon per day NEWater capacity at Tuas in the west of the island. It is also comprised of 40 kilometers of deep tunnel, some three to six meters in diameter and 60 kilometers of linked sewers, around 0.3 to three meters in diameter.[35] These large tunnel pipes are constructed using tunnel boring machines through the rock and mixed-face substrate of their routes through the island. Remote operated vehicles will be used for inspection and fiber optic cable will be embedded into the lining of the tunnels for monitoring of structural adequacy. With airflow also present in the tunnels, odor nuisance is mitigated through air management facilities located at junctions with the link sewers. Management of these engineering undertakings will use Building Information Modeling (BIM) to take advantage of coordination during construction and future asset management system. The accompanying map and diagrams present aspects of the system and the location of three treatment plant and NEWater reclamation facilities at Kranji in the north, Changi in the east and Tuas in the west. Integration of energy and water management approaches are also being intertwined with the DTSS, as described earlier.

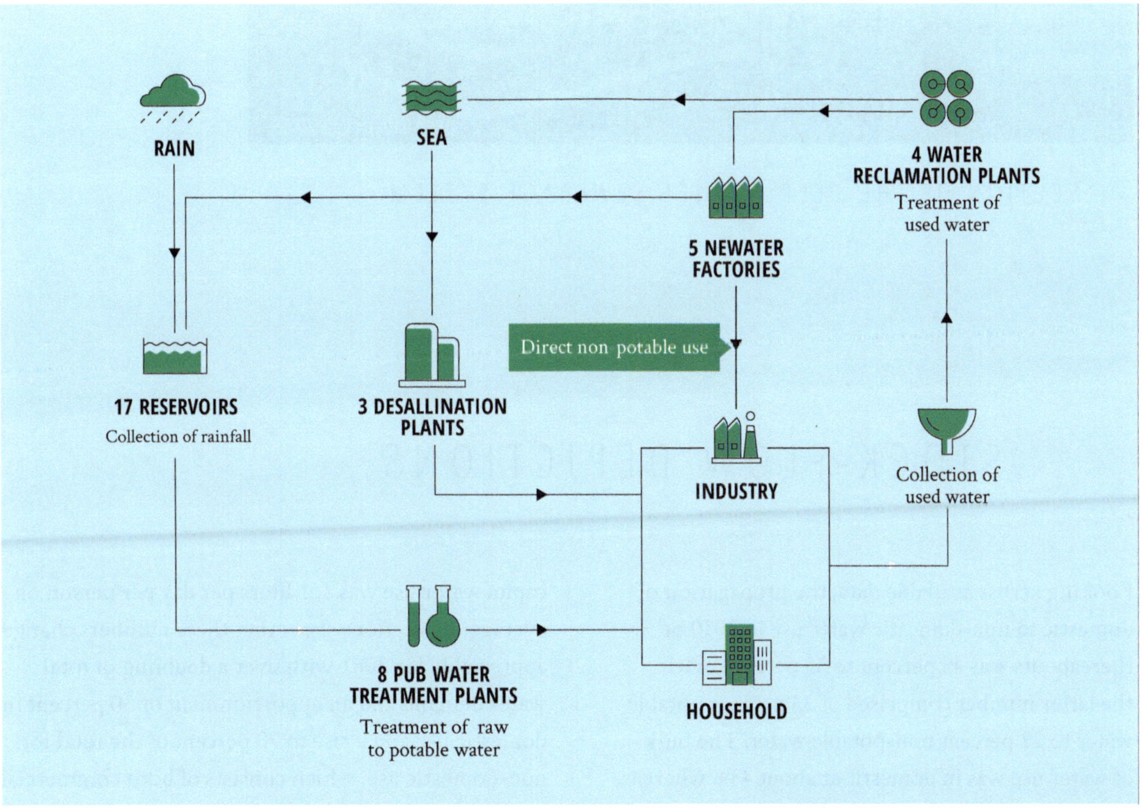

52. SINGAPORE'S WATER CLOSED LOOP SYSTEM

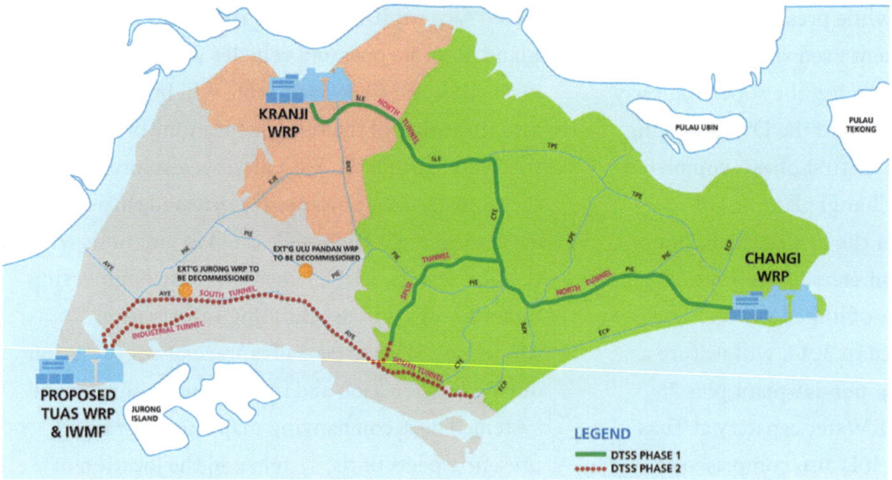

53. MAP OF THE DEEP TUNNEL SEWERAGE SYSTEM

54. SECTION OF THE DEEP TUNNEL SEWERAGE SYSTEM

d.

STOCK-FLOW DEPICTIONS

Looking across available data, the propagation of domestic to non-domestic water use in 2010 or thereabouts was 45 percent to 55 percent, with the latter number comprised of 33 percent potable water to 22 percent non-potable water. The bulk of water use was in domestic at about 43% whereas commercial and industrial each takes up about 26% of Singapore's water use. As noted earlier the per capita water use was 151 liters per day per person on average. Projections show that these numbers change appreciably for 2061 with over a doubling of total water demand and an apportionment of 30 percent in domestic use and a rise to 70 percent of the total for non-domestic use, which consists of both commercial and industry use.[36] The per capita use is estimated to be 130 liters per person per day, based essentially on a level likely to be reached by 2030. Returning to Singapore's four taps of sources of supply, in 2010 NEWater accounted for 19 percent of total use is also proportioned heavily in favor of industry, with only six percent for potable household use.[37] Water from desalination, accounting for 25 percent of total use is directed in the opposite direction, with 60 percent for household use and 20 percent each for industrial and building use. Water piped directly from Johor by sales accounted for 28 percent of use also again weighted strongly to household use at 54 percent, with 23 percent again evenly distributed for industrial and building use. The amount of unaccounted for water in this inventory is only five percent, a further testament to the closure of the water loop across the island.[38] Changes in these proportions in 2061, at least as projected, are also dramatic with the absence of piped water from Johor and Singapore aimed to increase the capacity of NEWater to meet up to 55% of the water demand, while up to 30% of the water demand will be met by desalination.[39]

Another way of accounting for water use and in relationship to other resources as well, like energy and land, is through what are called Sankey Diagrams. These diagrams, such as depicted here, are flow diagrams, with the width of areas shown in direct proportion to the flows of different quantities of resources, from initial raw inputs to the results of production in the form of industry, commercial and domestic uses in this case. These flow diagrams are named for an Irish captain by the name of Mathew Henry Phineas Riall Sankey who originally used this type of diagram to describe the steam engine and its relative efficiencies with regard to energy inputs in 1898. In a rather simple diagram he depicted a wide band of inputs on the left-hand side with output in the form of propulsion on the far right-hand side, and various forms of waste emanating from the process along the way, such as smoke, friction, alternator flare and massive amounts of condensation.[40] Another notable diagram of similar type was drawn up by Charles Minard in 1889, depicting Napoleon's Russian Campaign and actually pre-dating Sankey's diagram.[41] Nevertheless, these diagrams have been used subsequently, largely to represent energy and other basic resource inputs along with corresponding outputs and losses in efficiency or its equivalent. Unfortunately, at the time of this writing, there was not sufficiently accurate available data to construct a complete and meaningful Sankey diagram for Singapore across land, water and energy demand.

There was, however, availability and sufficiency of data to reflect dollar amounts expended by each use sector for its utilities, rather than literal physical usage. In other words, dollars expended becomes a proxy for physical consumption and a way of showing relative importance to the Singaporean economy. This kind of mixed stock-flow diagram also introduces the useful idea that resource flows are, in fact, tied to specific land uses and, more generally, show a relationship among utilities and land in a highly constrained Singaporean context. More specifically, this chart shows expenditures on water and electricity in Singapore, based on sales to commercial, industrial, residential, transport and other uses.[42] There is relatively balanced usage among all sectors, although it must be noted that similar expenditures does not necessarily equate to similar consumption. For example, in electricity sales, the difference in rate structure for each sector is substantial, so while commercial users consume fewer kilowatt-hours of electricity, the total expenditure by commercial users appears to be greater than that of industry. Electricity and water are routed to residential, commercial and industrial sites, which constitute only a portion of Singapore's land use. It is interesting to note that while a small share of land devoted to commercial buildings, their resource consumption is still significant, which speaks to dense commercial development patterns, Much more land is devoted to industrial and residential purposes. Though both utilities are critical, it turns out that electricity pricing is much higher, so that electricity sales dwarf water sales.

However, as projected in 2061, Singapore's water demand is likely to double, and projected demand for both industry and commercial would take up roughly 70% of the demand, with households taking up 30% of the demand. This could be a combination of the expansion of the industry and commercial sector in the future, while more effort needs to be put in place to encourage water conservation among households Nonetheless, PUB has consistently been able to achieve the target it has set to reduce the nation's water consumption per capita through public education and engagement programs for schools homes and commercial offices to encourage water conservation practices. In 2009, PUB also implemented the mandatory Water Efficiency Labelling Scheme, which requires manufacturers and retailers to inform consumers of the efficiency level of their products, such as washing machines, taps, urinal flush valves and so on. Unaccounted for water remains low at 5%, which is among the lowest globally. Dissipated energy is relatively high and correspondingly larger than unaccounted for

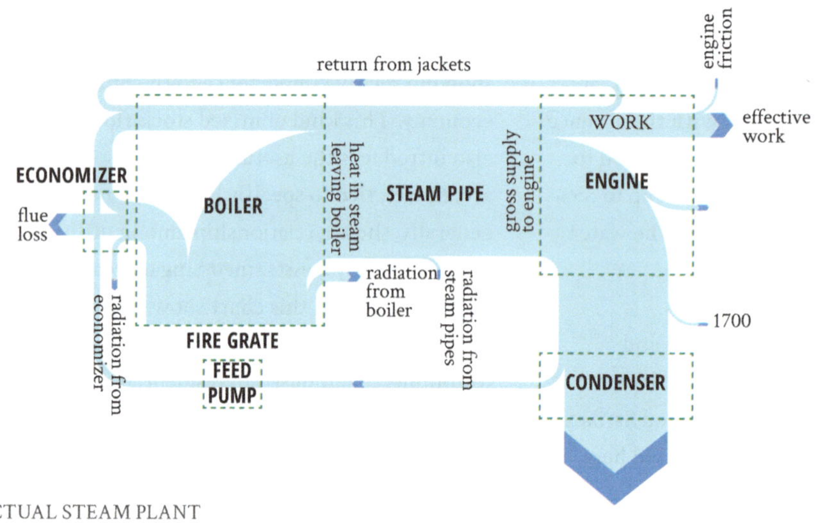

ACTUAL STEAM PLANT

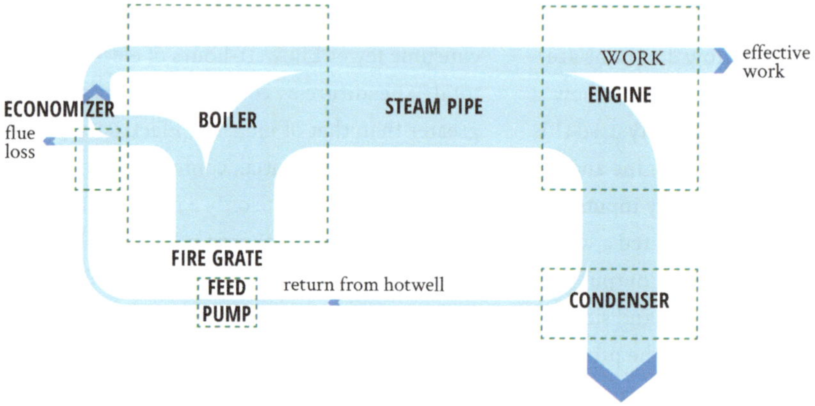

IDEALIZED STEAM PLANT

55. SANKEY DIAGRAM OF AN EARLY STEAM ENGINE

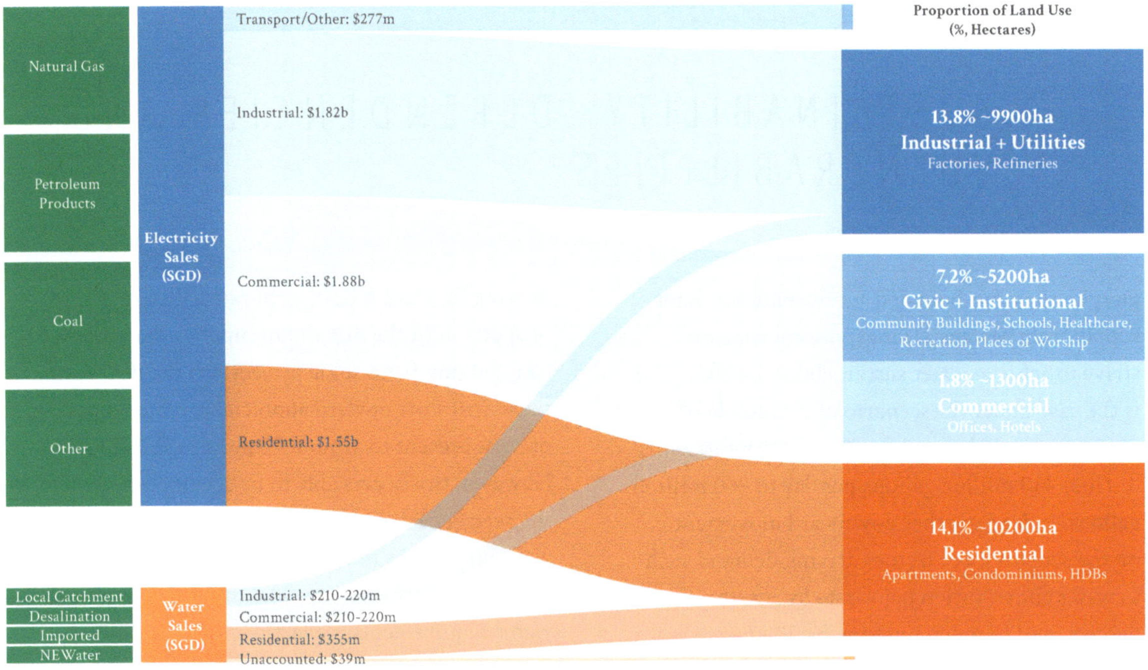

56. SINGAPORE'S ELECTRICITY, WATER AND LAND USE BY SECTORS (2017)

water. These proportions also reflect the city state nature of Singapore as an object of analysis. After all, with the land area for residential, commercial and industrial development taking up only about 36.9%, it is important to note that Singapore is a land-scarce island without a natural hinterland. Therefore, the city-state also has to allocate land for other national uses such as for defence, national port and airport, nature reserves, parks and reservoirs. Many of these land uses are beyond the scope of this diagram yet are critical in ensuring a balanced ecosystem and sustainable environment, which therefore has a consequential impact on Singapore's water supply especially within the local water catchment.

Outside of the internal representation of the national taps, total water demand including virtual water associated with the importing of food from crops and livestock, as well as from industrial production must also be incorporated into the consumption picture. One estimate places the total net virtual water import figure at 11,781 million cubic meters of water per year and across a period from 1987 to 2001. This compares to a demand within the island of around 430 million gallons per day, or roughly 766.8 million cubic meters per year, a relatively small proportion of the grand total including the virtual water volume. Of this the bulk is again in industry and industrial products with only 0.03 percent of the net virtual water importation equivalent being consumed on the island by livestock and crops, or the basic food component.[43] This might also seem to be something of a justification to move away from local food production in the 1980s described earlier, in favor of the industrial and service sectors and with much higher added value of production. Nevertheless, these figures also illustrate the extent to which Singapore's effort to reach water sustainability in the domestic sector is both localized and of strategic importance. It also illustrates one of the trade-offs in pursuing such a strategy within the broader international scheme of things.

e.

SUSTAINABILITY, DEPENDENCIES AND VULNERABILITIES

Despite the time-bounded agreements for water supply from Malaysia, Singapore continues to strive to achieve water sustainability by 2061. This is the case under the scenario of a doubling of expected water demand from the present to 2061, or from 430 million gallons per day to 860 million gallons. Indeed, further downward movement on the sensitivity of factors driving demand such as water conservation per capita by five to seven percent, and island population by 20 percent and predominantly among non-residents, could lessen the overall demand to 640 million gallons per day and within projected downward shifts in the proportion of domestic water use to 30 percent by 2061, domestic demand by households would remain at current levels of around 194 million gallons per day.[44] Any water supply scenario moving forward will incorporate a different proportion in the mix of tap supplies than today. In fact, judging from some projections the degree of reuse will shift upward appreciably from 44 percent or so at present to around 72 percent in 2061. However, this seems able to be accomplished under the reference designs for NEWater and desalination plant operations in the future, including likely technical improvements. The inclusion of virtual water requirements, nevertheless, alters the rosy picture appreciably with an estimated 11,781 million cubic meters per year of virtual water added to the current 767 million cubic meters per year of island water demand.

Major among the dependencies of water with other sources and stocks of supply are energy

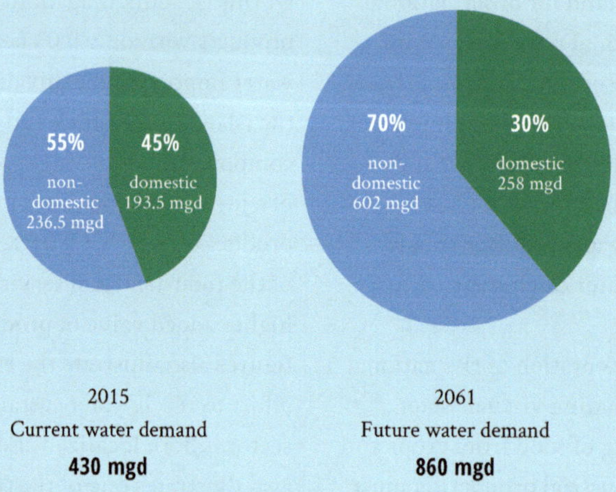

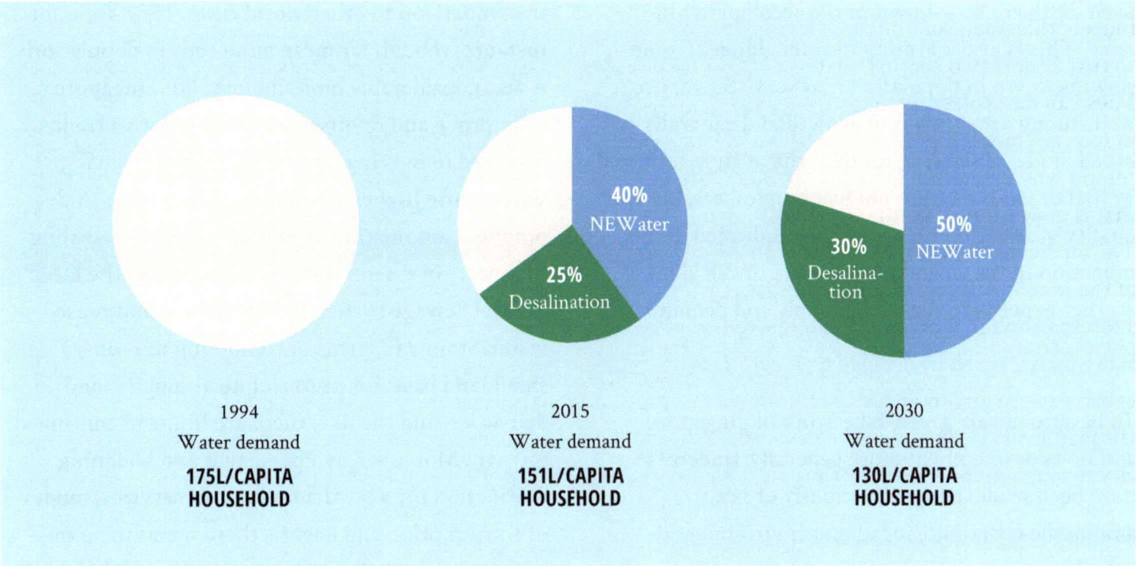

57. SINGAPORE'S WATER DEMAND AND SUPPLY STRATEGY, 2015 AND 2030

use, population and food, as well as industrial products in the case of virtual water demand. Energy demand, though seemingly a real issue during the early days of non-conventional water treatment, has abated to levels slightly above current practice, but not alarmingly so, with newer technologies. The replacement of highly energy-consumptive reverse osmosis processes with newer electrodialysis-electroionisation techniques puts treated reused water in the range of 1.8 kwh per cubic meter of water and 0.75 to 0.95 kwh per cubic meter overall.[45] Without any abundance of local energy supplies the major trade-off is land base necessary for storage, refining and production facilities, as well as any residuals of disposal of this production in the form of air and water pollution. On a small island it might be argued that such facility presents one an issue of needless competition with higher-value uses. However, so far, this does not appear to be the case. In fact, piece-meal technical advancements, such as the Deep Tunnel Sewage System work in the opposite direction both with regard to energy use and land conservation.

Population growth in Singapore has been advancing at around 2.0 to 2.5 percent per annum, standing now at 5.7 million or so inhabitants. Continuation of the overall and espoused growth policy will require around 15,000 to 25,000 new citizens each year, particularly as the hoped for 2.1 Total Fertility Rate (TFR) has never been reached with TFR today standing at only 1.2. Furthermore, the aging population, especially of 'baby boomers' is at a cross-roads, with a further 900,000 people expected to exceed 65 years of age before 2030.[46] One upshot will be the continued need for in-migration in order to maintain levels of social welfare, if not material well-being. Another outcome, however, might also be some containment of this in-migration particularly of non-residents in Singapore because of other strains placed on the welfare of the city-state by way of physical development, housing, congestion and basic supplies of resources like energy, water and land. Currently Singapore, with around 8,041 persons per square kilometer, is the second sovereign state in density, just behind Monaco. The likely proportion of non-resident population is estimated by 2030 to be on the order of 45 percent, and has been on the rise as shown in the accompanying table. This is also alarming to some. Indeed, some slowing down in population may serve Singapore well among other forms of growth and generally fit a broad model of striving for quantity at first, followed by higher consideration not just of numbers, but the quality of added contributions. As indicated earlier, reduction in the role of population growth alone accrues benefits to the water supply and demand perspective.

To be sure in this 'green-blue' story of Singapore and in its development more generally, tradeoffs have been made, either consciously or not, among the economic, social and environmental interdependencies that constitute and, in fact, describe and define the island state. As noted earlier, Singapore is not renowned for any particular brand of goods although it is a popular destination for business, consumption and tourism. It has a highly qualified work force and society, but its economic intensity is moderate to high, although not very high in comparison to other world cities. New York, for instance, though far more numerous in population is also considerably more intense. For Singapore to improve and to offset potential negative trends involved in moving away from growth policies, will require higher efficiencies, innovation, and potential automation. It may also require rescaling of aspects of essential services, much like the Deep Tunnel Sewage System has for treated and reused water. Simply in terms of competing uses on a small land base, limits to technical supplies and deliveries, and the like, there are limits to continued growth. Moreover, as a persistent and widening destination for a particular band of services, modes of consumption and leisure, there seems to be the very real potential of falling into what might be described as a 'high-to-mid-level trap', where further significant advancement is limited. Such an outcome, however, may not be such a bad thing, but it is one in which the very strong and profound influence of water sustainability and the environment on the national narrative have come into play.

NET VIRTUAL WATER IMPORT (1997–2001)	TOTAL (104 M3/YEAR)	PER CAPITA (L/P/DAY)
CROP PRODUCTS	2,386	1,634
LIVESTOCK PRODUCTS	1,461	1,000
MEDICAL PRODUCTS	7,934	5,435
TOTAL	11,781	8,069

58. SINGAPORE'S VIRTUAL WATER USE

A related potential vulnerability is in the domain of virtual water consumption and what it implies by way of other dependencies, although responsible consumption is probably more apt here as a basic concept. With regard to food, crops such as wheat and rice, can be distinguished from livestock, including chickens, pork and eggs in Singapore's case. As it turns out Singapore imports a substantial amount of wheat from Australia which uses two times more water on average than another source like the United States. Most rice in Singapore is also purchased from Thailand where the virtual water content is also high, although it is less productive per unit of crop production than non-exporting China and India.[47] Total virtual water consumption for livestock is significantly lower than for crops, at 1,461 million cubic meters of water per year compared to 2,386 million cubic meters. Also diet plays a role here with substantial amounts of chicken and pork being imported, although with a consumption of eggs that is about one tenth that of Europe. With regard to industrial products, by far the largest consumer of virtual water content at 7,934 million cubic meters of water per year, on average, different trading partners have different levels of virtual water content with the United States, for instance, at 100 liters per US$, compared to Germany and the Netherlands at 50 liters per US$ and a broad range for other countries from 10 to 151 liters per US$. Construction materials, mostly imported into Singapore, are relatively high in virtual water content and cotton, used in clothing, is extraordinarily high. In general there are two approaches that can be taken towards responsible consumption of virtual water. The first would be striving to achieve high levels of food self-sufficiency. As narrated earlier, this was broken off in the 1980s in order to pursue a 'clean and green' Singapore as much as anything. Prior to that the nation was self-sufficient in chicken, pork and eggs, as noted. Given the failure of attempts to pursue high-tech agriculture in Singapore, the likelihood of successful further pursuit of self-sufficiency seems to be low, although not impossible, given some less land-consumptive forms of vertical agriculture. A second broad strategy is to introduce virtual water content into consideration in trading agreements with various nations for foodstuffs and products. Some other existential threats to Singapore's blue-green picture may also be seen to come from climate change, sea level rise and other weather perturbations. However these will be dealt with in a later chapter.

endnotes

1. Oon, C. 2009. Key step to water adequacy. *Straits Times*, June 24

2. Lenouvel, Vincent; Michel Lafforgue; Catherine Chevauche and Pauline Rhétoré. 2014. "the energy cost of water independence: the case of Singapore" *Water Science and Technology: a journal of the International Association on Water Pollution Resources*. 70:5, p.789.

3. Lenouvel et al., p.789.

4. Interview with Mr. Harry Seah, Director of Technology of PUB on 6 February 2017.

5. Tan, Irene. 2016. *Beyond Singapore's blue-green matrix: an international comparison of blue-green systems in Melbourne and Singapore* (Unpublished paper, 14 July).

6. The Public Utilities Board was set up as a statutory board in 1963 to provide electricity, piped gas and water supply for Singapore. In 2001, the electricity and gas undertakings of the Public Utilities Board were transferred to a newly set up statutory board called the Energy Market Authority. The Public Utilities Board was then reconstituted as the national water agency, known officially as PUB, Singapore's National Water Agency. The idea was for PUB to manage Singapore's water supply, water catchment and used water in an integrated and more holistic manner.

   Based on an interview with Mr. Tan Gee Paw, Chairman, PUB on 9 February 2017 and Channel News Asia. "PUB to expand groundwater monitoring network" (26 May 2016, 11:23 am).

7. Lenouvel et al., p.789.

8. Lenouvel et al., p.787.

9. Allan, J. A. 1998 (2005) "Virtual Water: A Strategic Resource, Global Solutions to Regional Deficits". *Groundwater*. December, pp. 545-546.

10. Hoekstra, A. Y. and A. K. Chapagain. 2007. "Water Footprints of Nations: water use by people as a function of their consumption patterns". *Water Resources Management* 21:1, pp. 35-48.

11. Vanham, D. 2011. "How much water do we really use? A case study of the city-state of Singapore. *Water Science and Technology-Water Supply*. 11:2, p.234.

12. Lenouvel et al., 2013, p789.

13. Based on assumption of author in Lenouvel, p.789.

14. Lenouvel et al., p.789.

15. Microfiltration is a type of physical filtration process where a contaminated fluid is passed through a special pore-sized membrane to separate micro-organisms and suspended particles from process liquid.

16. PUB, "Innovation in Water Singapore", June 2011, https://www.pub.gov.sg/Documents/InnovationWater_vol1.pdf

17. Water: From Scarce Resource to National Asset, Urban Systems Studies Booklet Series. Singapore: Cengage Learning Asia, 2012.

18. Lenouvel et al., p.786f.

19. Lenouvel et al., 2013, p789.

20. Vanham, P.269.

21. Lenouvel et al., 2014.

22. PUB, 2016.

23. Vanham, p.289.

24. Seah, Harry; K. L. Khoo; J. Y. Chua;mD. Toh and S. C. Chua. 2010."Cost effective way to harvest estuarine water: variable salinity desalination concept" *Aqua – Journal of Water Supply: Research and Technology* 59:6-7, pp.453-454.

25. Seah, p.458.

26. Machuca, Lubomir and Vit Fara. 2014, "Combination of Electrodialysis and Electrodeionization for Treatment of Condensate from Ammonium Nitrate Production." *World Academy of Science, Engineering and Technology International Journal of Energy and Power Engineering*, vol 8, no 6, pp. 485-487.

27. Siemens. 2017. *Ingenuity for Life: Our History in Singapore.* https://www.siemens.com/ and "Siemens to develop innovative sea water desalination technology" (Monday, June 23, 2008).

28. Based on an interview with Ms. Olivia Lum, Founder of Hyflux on 15 August 2017.

29. Public Utilities Board. 2016. *Our Water, Our Future.* (Singapore: Public Utilities Board).

30. Interview with Ms. Linda De Mello, Deputy Director, 3P Networks, PUB on 16 August 2017.

31. Interview with Mr. Harry Seah, 6 Feb 2017.

32. Seah, p.452-458.

33. Lenouvel et al., p.792.

34. Based on an interview with Mr. Yong Wei Hin, Director of the Deep Tunnel Sewerage System Department, PUB, on 22 June 2017.

35. Interview with Mr. Hin.

36. See data in accompanying tables.

37. Lemonick, S. Drinking toilet water: The science (and psychology) of wastewater recycling. *Earth Magazine*, [online]. Available at https://www.earthmagazine.org/article/drinking-toilet-water-science-and-psychology-wastewater-recycling.

38. Remainder of the paragraph based on Vanham, p, 222.

39. Vanham, 2011 (p22).

40. "Sankey Diagrams",https://en.wikipedia.org/wiki/Sankey_Diagram. Also see, Schmidt, Mario. 2008. "The Sankey Diagram in Energy and Material Flow Management, Part 1: History" *Journal of Industrial Ecology* 12:1, pp. 82-94.

41. Corbett, John. 2001. *Charles Joseph Minard: Mapping Napoleon's March, 1861.*(Washington, DC: Centre for Spatially Integrated Social Science).

42. 43% of water is estimated to be sold to households; 5% of water is unaccounted for; and the remainder is split among industrial and commercial users. We used domestic and non-domestic water prices for these two categories. Based on PUB's recommendation, we treat water sold for resupplying ships, though non-zero, as negligible. Annual sales total around $785 million. We assume that unaccounted-for water (which may include leaks, flushing or other losses) is equally spread throughout the system, so if that water were to be sold, it might be worth around $39 million. The 5% of water unaccounted for in the system reflects an extremely low rate of lass by global standards. For electricity, in reality, there are both contestable and non-contestable electricity rates, with different shares of each sector subscribed to different regimes. Here, we have used the assumption that no residential households subscribe to conetstable electricity rates, but the majority of industrial and commercial use contestable pricing. Estimates for contestable expenditures relies on a simplified price based on Uniform Singapore Energy Price (USEP). Our electricity data is based on the twelve months of 2017. By the end of 2018, all of Singapore will be an open electricity market, which will also yield new pricing outcomes in the future.

43. Based on Vanham (2011), Allan (1998), Hoekstra and Chapagain (2007) and PUB (2016).

44. PUB. 2016. *Our Water, Our Future.* (Singapore: PUB).

45. H2PC Asia, n.d. Energy efficient seawater desalination in Singapore. Available at http://www.e2singapore.gov.sg/DATA/0/docs/NewsFiles/Energy%20efficient%20desalination.pdf

46. See Department of Statistics Singapore, *Population Trends*, 2017.

47. Paragraph based partially on Vanham, p.227 and p.45f.

chapter

05

GARDENS, PARKS AND GREEN RESERVES

P. G. Rowe and L. Hee, *A City in Blue and Green*,
https://doi.org/10.1007/978-981-13-9597-0_5

Dating well back into the colonial period, Singapore began to make a mark in the botanical, horticultural and wildlife worlds. This came by way of both its natural circumstances, including those that survived the onslaught of plantation and other less than discriminating agricultural practices and in the form of its magnificent Botanic Gardens and other related curatorial activities. Biodiversity is also a hallmark of Singapore's environment, although it was significantly depleted during the colonial era of land exploitation. In the shift that has occurred towards a 'city with or in nature', significant aspects of Singapore's plant and wildlife attributes have been pushed further into service. This shift also appears to expert opinion to be closer to a truer natural state of existence than earlier garden-like interpretations, even though the appropriation of whole-hearted public support might be more difficult to secure. Depending upon the eye of the beholder, it certainly seems true that butterflies are more beautiful than caterpillars, but, of course, without the latter you will not have the former. More prosaically, the greening of roadway verges, public parks and so on, thrive better under mixed rather than mono-cultural species conditions. Also the scale of vegetation involved can be and often is very mature and large, more in keeping with the primeval tropical antecedent conditions and so-called 'nature' of Singapore. For this to be sustained successfully, risks from events like falling branches and uprooted trees must be avoided almost at all costs in Singapore's current socio-cultural environment. Consequently the National Parks Board has begun engaging in high-tech data applications to these environments. Finite element analysis, for example, usually deployed on building structures is now routinely applied to large trees to help predict branch failures and to monitor trimming and other maintenance functions. Similar data rich scrutiny of ways of improving biodiversity and wildlife habitat are under way, including fluid dynamic modeling of water ways, especially in vulnerable coastal areas. The PUB's ABC Waters approach to public parks, like the Kallang River at Bishan-Ang Mo Kio Park, and to drainage canals, like the Alexandra Canal, are aimed squarely at closer integration of a more 'natural' agenda of greening and conservation than the earlier strictly structural solutions to conditions like storm water management. A more complete embrace of local tropical landscapes, however, remains to come. Other significant projects, like 'Gardens by the Bay,' while attracting substantial local and tourist attention, also thrusts the curatorial program of botanical activities into the twenty-first century. The remarkable hybridization and DNA protocols of researchers at the Singapore Botanic Gardens also extends Singapore's reach into the forefront of tropical plant life and ecology, as does the literal greening of buildings that has sprouted forth in the last decade or so. These mergers of science, technology and active urban environmental management are unique and, again, appear to be potentially transferable elsewhere. Over time, Singapore appears to be moving towards a distinctly biophilic outcome.

a.

BOTANIC AND HORTICULTURAL GARDENS

Specific types of gardens have existed in human history for many years. In the western ecumen, for instance, this interest by the ancient Romans first started to focus on the medicinal properties and later, around the eighth century, monks brought this interest forth in their monastic gardens.[1] The rise of the botanic garden, however, began with the sixteenth-century Italian university gardens, such as the very first in Europe at the Universita di Pisa, founded in 1543 by Luca Ghini both a physician and a botanist. By the mid-eighteenth century and later, the botanic garden was essentially a museum of living plants. Like other kinds of museums it had a double function.[2] It was a place where plants could be studied by experts for the furtherance of scientific knowledge and a place where exhibits could be arranged for the education and recreation and enjoyment of non-experts. At much the same time international trade propelled imperial powers to bring newly discovered tropical species to Europe and to cultivate them both in the homeland and in their natural environment. Established in 1859, Singapore Botanic Gardens served as a park for Singaporeans and visitors, a scientific institution and as a testing ground for tropical plantation crops. The first of these functions was mainly for the Singaporean elite at the time. In the 1800s it was a centre for research and plant conservation with a focus on economic botany that had direct repercussions, as pointed out earlier, on Malaya and Singapore. Also, among other characteristics, the Botanic Gardens was the largest intact historically-designed landscape in Singapore. It is also among the most visited botanic gardens in the world, hosting as many as four million visitors in 2013. Then, the Singapore Botanic Gardens was inscribed as a UNESCO World Heritage Site on 4 July 2015 at the 39th session of the World Heritage Committee, the very first of its kind in the island nation.

59. SINGAPORE BOTANIC GARDENS

Historically, after being founded in 1859 on a site at Tanglin, at least partly under the banner of the Gardens of the Agri-Horticultural Society of Singapore, Lawrence Niven was engaged in 1860 to develop the 23 or so hectare property as a garden.[3] This he did in the manner of the English Landscape Movement and of 'Capability' Lancelot Brown, and by 1864 a system of roads and footpaths was in place. Then, in 1866, the Society's finances permitted expansion to the northwest, including the Swan Lake. However, having suffered substantial cost overruns during the construction of a Superintendent's house, the Agri-Horticultural Society appealed in 1874 to the government for relief. Henry James Murton was then appointed in 1875 as the Gardens Superintendent, with the assistance of the Royal Botanic Gardens at Kew in Great Britain. Collecting plant materials from Malaya and Sri-Lanka, among other places, he turned the gardens into a more typical botanic garden with a focus on plants of economic interest. In 1879, for instance, he oversaw the construction of the spectacular 'Palm Valley'. He also established a zoological collection that was terminated in 1904. Nathaniel Cantley succeeded Murton in 1880 and who set about to rationalize buildings within the gardens. A herbarium was built, for instance, in 1882 and nurseries for plant trees was established. Both Murton and Cantley were trained at Kew. An avid botanist Cantley also placed the Gardens on a firm systematic footing.

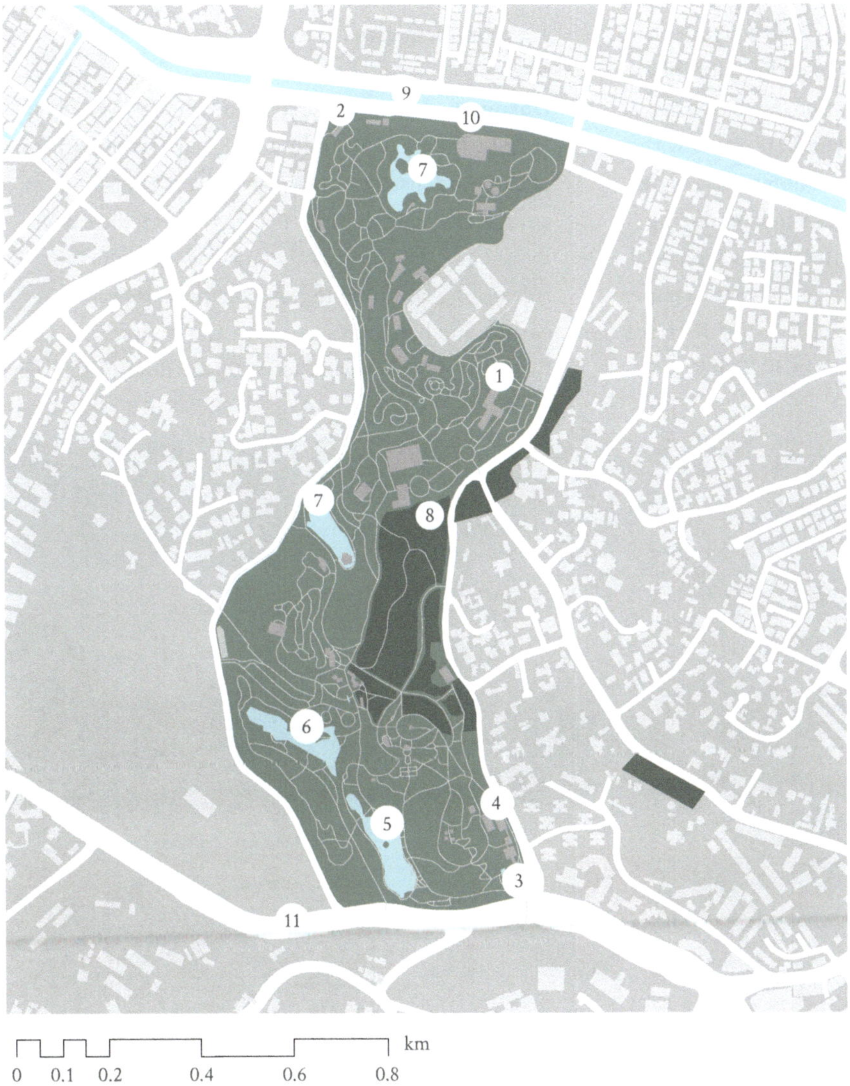

1. Raffles Building
2. Bukit Timah Gate
3. Tanglin Gate
4. Botany Centre
5. Swan Lake
6. Symphony Lake
7. Eco Lake
8. Forested Area
9. Rochor River
10. Bukit Timah Road
11. Holland Road

60. PLAN OF THE SINGAPORE BOTANIC GARDENS

Henry Nicholas Ridley took over from Cantley in 1897, becoming the Garden's first 'Director' and making it a regional centre for understanding the flora of Southeast Asia, a position it retains today.[4] Ridley's interests were broad, encompassing zoology, geology and botany with a special interest in the latter category for orchids. Coming from the National History Museum in London, he launched the first scientific agricultural journal in the region – The Bulletin of the Malay Peninsula – and became involved, as noted earlier, in rubber plant development and cultivation. Issac Henry Bukhill took over from Ridley in 1912 and saw the Gardens through a critical transition period from British rule to part of Malaya. Much later during the Fourth World Orchid Conference in 1963 held for the first time in Singapore, he delivered a paper on Singapore's role in hybrid propagation. Richard Eric Holttum succeeded Bukhill in 1925. He was a scientist and emphasized horticulture with orchids as a focus.[5] As early as 1928 he set up an *in vitro* propagation unit, which from the 1930s onwards began to see the merits of this approach, resulting in 1956 with the beginnings of the Gardens VIP Orchid Nursery Program. Certainly by the mid -1950s the Gardens was known globally for orchid propagation. Today staff's interest and work in genetic aspects of plants and DNA sequencing for plant identification are prominent areas of expertise.[6]

During the Japanese occupation from 1942 to 1945, Hidezo Tamakadate and Kwan Koriba were Directors of the Gardens. After the war Murray Hudson Ross, curator of the herbarium before the war, took over. Eventually the Gardens then played an important role during the greening of Singapore described earlier, and being taken over by the National Parks Board in 1986. During the 1970s the Botanic Gardens

61. SINGAPORE BOTANIC GARDENS

also assumed the role of a public park, now 65 hectares in area.[7] Today's Gardens are divided into three main cores: Tanglin, Central and Bukit Timah. The heritage portions are located in Tanglin; the tourist belt is located in the central area, including the Orchid Garden, the Healing Garden and the Fragrance Garden. The Eco-Lake and Foliage Garden are at Bukit Timah, which is an educational and discovery zone. One of the most, if not the most, important centres of taxonomic and biodiversity research in the region, in 2014 the Gardens had 36,400 living plant accessions, 6,500 species and 44 heritage trees, a herbarium of 750,000 species of which 8,000 are typical specimens. In addition there is a library of over 28,500 books, journals and unpublished data.[8]

1. Flower Dome
2. Cloud Forest
3. Malay Garden
4. Chinese Garden
5. Indian Garden
6. Colonial Garden
7. Secret Life of Trees
8. World of Palms
9. Understorey
10. Fruits and Flowers
11. Web of Life
12. Discovery
13. Supertree Grove
14. Golden Garden
15. Bay East Garden: Planned Location of the Founder's Garden
16. Marina Barrage
17. Marina Bay Sands

62. PLAN OF GARDENS BY THE BAY

The second extraordinary botanical and horticultural contribution by Singapore is the Gardens by the Bay project, first announced in 2005 by then Prime Minister Lee Hsien Loong on its 101 hectare site adjacent to the Marina Reservoir in central Singapore. It was the subject of an international competition in 2006 won by Grant Associates for the Bay South Garden and Dominic White for the Bay East Garden. The third garden – the Bay Central Garden – is a link between the other two gardens. A central and prominent feature of the Gardens by the Bay is the two conservatories beside the Marina Reservoir.[9] Both were designed by William Eyre to be energy efficient. The Flower Dome is the larger of the two and sits on a 1.2 hectare site and rises 38 meters. It is the largest column-less greenhouse in the world and maintains an interior atmosphere of between 23 and 25 degrees centigrade, replicating a mild dry climate and featuring plants found in the Mediterranean area, as well as in semi-arid and tropical regions like Australia, South America and South Africa. It is comprised of seven different gardens which also merge together in unique ways as a singular display across a sloping topography.[10] The Cloud Forest is the second conservatory. It occupies a 0.8 hectare site and rises to enclose a 42 meter tall cloud mountain, accessible by elevator and covered in epiphytes such as: orchids, ferns, mosses and ormeliads. An exterior catwalk allows visitors to descend around the mountain's edge to the lower entry level. In addition, another prominent constructed feature is the super-trees that are from 25 to 50 meters in height.[11] They are outfitted with environmental technologies, like photo-voltaic cell and rainwater devices, which allow them to mimic ecological functions. They are also environmental engines for the gardens, having hot air evacuation units and methods of cooling water.

63. THE FLOWER DOME OF GARDENS BY THE BAY

64. THE CLOUD FOREST OF GARDENS BY THE BAY

65. THE SUPERTREES OF GARDENS BY THE BAY

66. GENERAL VIEWS OF GARDENS BY THE BAY

The overall site is further comprised of roads and pathways defining varying horticultural domains, ranging from open meadows to dense forested areas. Part of this arrangement are the heritage gardens under the 'Plants and People' theme, representing plants from Chinese, Malay and Indian circumstances. The other larger part is under the 'Plants to Plant' theme and displays connections and relationships involving plant diversity. The entire ensemble is replete with rocks and other materials gathered on periodic overseas trips by the garden's curatorial staff, as well as several striking sculptures interspersed among the plants and accompanying water bodies. Another area is undergoing development on the eastern side of the Marina Reservoir and across the Barrage emplacement, which will house a Founders' Memorial Garden.[12] The entire complex occupies reclaimed land that lay fallow for 35 or so years. There were early plans for its development. In fact, in 1984 Kenzo Tange and I.M. Pei were commissioned to propose layouts and designs for a new downtown on 266 hectares of the Marina South site. Tange, taking up the greening vision of Singapore proposed a radial scheme of intense development separated by swaths of green areas, drawing on the character of Singapore as a tropical island. Pei proposed a grid model that was integrated with the existing Central Business District. This facilitated the sales of land parcels incrementally. Commissioned at the behest of Lee Kuan Yew, the government adopted Pei's model which shaped the subsequent concepts of Marina Bay development.[13] Finally, the Gardens by the Bay was opened in 2012, attracting some 6.4 million visitors in 2014. It was built with budget and operations for around $58 million per year. It is without doubt one of the most extraordinary botanical installations in the world and chiefly the brainchild of the congenial and brilliant Dr. Tan Wee Kiat, an eminent plant expert, especially with regard to orchids.

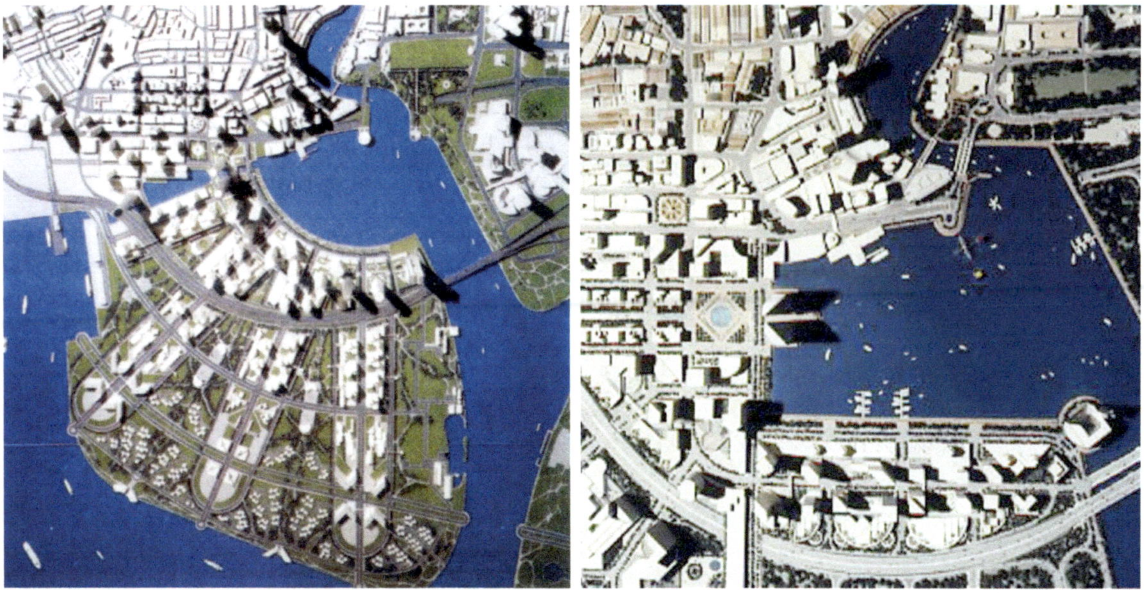

67. MARINA SOUTH STUDY 1983, BY KENZO TANGE (LEFT) AND I.M. PEI (RIGHT)

b.

NATURE RESERVES

As described earlier, during the last third of the nineteenth century, following on the heels of extensive deforestation for agricultural purposes, the extent and scope of mature species and forest trees came under scrutiny with the McNair Report of 1879. This was followed in 1884 with the creation of the Forestry Department in order to conserve remaining forested assets on the island, totaling, as noted earlier, around eight percent of Singapore's total area. The Municipal Catchment Reserve was established in 1900, along with several other reserves at Bukit Timah and some mangrove reserves, especially the Pandan Forest Reserve. Within reserves, reservoirs were constructed as noted in order to serve expanding population growth. MacRitchie was created in 1867, Kallang in 1911, and Sungei Seletar in 1922 with expansions in 1967 and 1969. Generally, early colonial administration of Singapore had little land management. Unfortunately, in the end, natural reserves of dryland primary forest amounts to only 0.5 percent of its original area.[14]

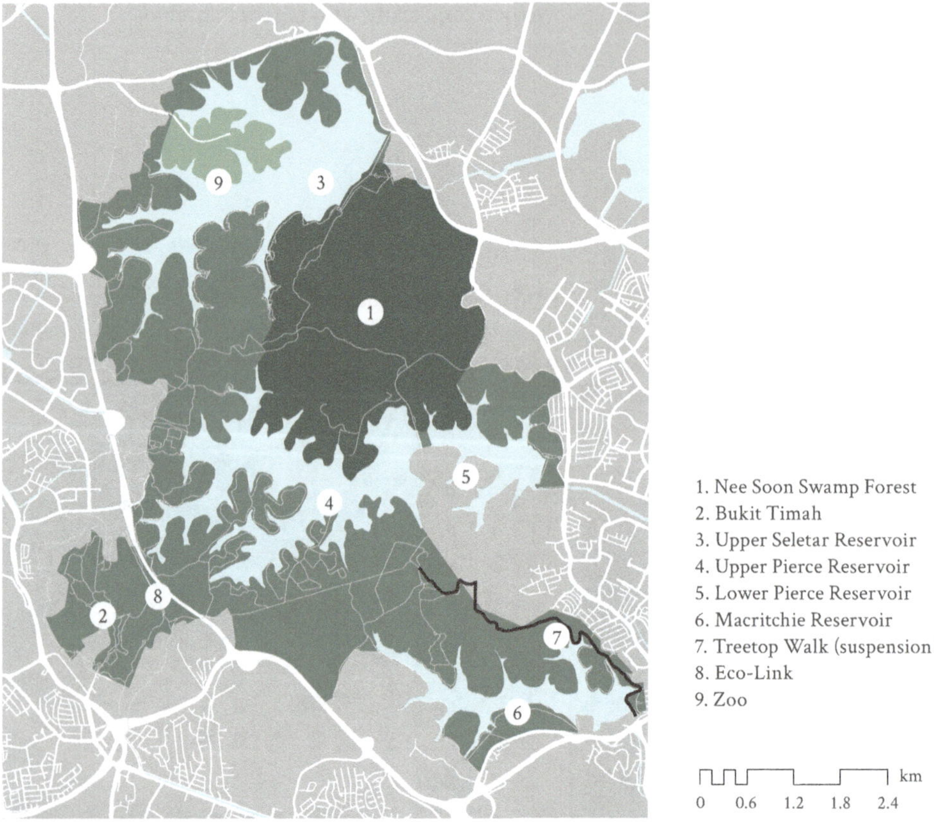

68. THE CENTRAL CATCHMENT AND THE BUKIT TIMAH NATURE RESERVES

Among Singapore's four major nature reserves there is the Central Catchment Nature Reserve, the Bukit Timah Nature Reserve, the Sungei Buloh Wetland Reserve and the Labrador Nature Reserve.[15] The Central Catchment Nature Reserve is the largest, spanning 2800 hectares in area and acting as a large green lung in the centre of Singapore. Indeed, it is an area ringed by the 1971 Concept Plan and is home to patches of dipterocarp forests, the species of rich primary lowland forest that was once characteristic of the island. It also hosts the rarer primary freshwater swamp forest, such as the Nee Soon Swamp Forest. The Central Catchment is the major one of only two such catchments, earning its name from housing the MacRitchie, Upper Seletar, Upper Pierce and Lower Pierce Reservoirs. A primary site for hunting, bird and nature watching, the MacRitchie Reservoir area, for instance, comprises 20 kilometers of trails and boardwalks. There is also the spectacular 250 meter long suspension bridge and observation tower, offering unrivaled high-level views of the forested area. These days the Central Catchment Nature Reserve is linked across the Bukit Timah Expressway by the Eco-Link@BKE to Bukit Timah.[16] This link is the first of its kind in Southeast Asia and aims to restore ecological connection between the two adjacent nature reserves, allowing some wildlife to expand their natural genetic pool and survival chances. The two reserves combined comprise some 840 or more species of flowering plants and some 500 species of fauna. The eco-link is hour-shaped in plan with trees and shrubs planted on the bridge. A wire mesh fence across the Central Catchment side of the link discourages larger animal species access to the much smaller habitat of the Bukit Timah Nature Reserve.[17]

69. THE ECO-LINK @ BUKIT TIMAH EXPRESSWAY

70. THE TREETOP WALK AND SUSPENSION BRIDGE IN THE CENTRAL CATCHMENT NATURE RESERVE

The Bukit Timah Nature Reserve is located 12 kilometers from the centre of urban Singapore and is a very rich and diverse ecological system.[18] It is about 164 hectares in area and the highest point in Singapore at 163.63 meters above sea level. Originally established in 1883 it is one of the very few areas of primary rainforest in the country. It has also been a botanical collection area for over 100 years and is reputed to have approximately 40 percent of the nation's flora and fauna.[19] 1951 saw island-wide enactment of a Nature Reserves Ordinance, along with establishment of a Nature Reserves Board to manage these areas. In 1990 both the Central Catchment Nature Reserve and the Bukit Timah Nature Reserve were gazetted for propagation, protection, and preservation of Singapore's flora and fauna. This was also reinforced by the Parks and Trees Act of 2005. Then, Sungei Buloh Wetland Reserve was declared an ASEAN Heritage Park in 2003, followed by Bukit Timah Nature Reserved which was declared in 2011. They both became part of the prestigious network of 35 protected areas within the ASEAN member states. Like the Central Catchment Nature Reserve, Bukit Timah Nature Reserve plays host to trekking, mountain biking and, consequently, falls under Singapore's rules of trail etiquette.

The two smaller reserves are the Sungei Buloh Wetland Reserve at 130 hectares in area and the Labrador Nature Reserve at ten hectares in area.[20] The Sungei Buloh Wetland Reserve is located in the northwest of the island and was gazetted as a nature reserve in 2002. As mentioned earlier, it is replete with a rich ecology of plant and animal life. The Labrador Nature Reserve, by contrast, is located on the southern edge of the island, facing out to sea and is part of the Southern Ridges built on the edge of a secondary forest with cliff-side vegetation and picturesque views. It was also gazetted in 2002 as a nature reserve. The Southern Ridges is a topographically interesting area of Singapore comprising Mount Faber Park, Telok Blangah Hill Park, and Kent Ridge Park. Collectively, the four nature reserves make up most of the natural vegetated landscape in Singapore, along with the most numerous and diverse species of flora and fauna. When the metaphor of a 'City in Nature' is used it generally refers to this kind of landscape and aesthetic.

71. THE SUNGEI BULOH WETLAND RESERVE

72. LABRADOR NATURE RESERVE

C.

THE ACTIVE, BEAUTIFUL, CLEAN WATERS PROGRAMME

The Active, Beautiful, Clean Waters (ABC Waters) Program was implemented in 2006 as a part of Singapore's storm-water management strategy and reflects the nation's move towards 'blue-green' sensitive urbanization through the adoption of a low-impact development ideology and practices.[21] In this manner the ABC Waters Programme seeks to transform the utilitarian drains, canals and reservoirs, which have served the island from colonial days onwards into beautiful and clean waterscapes which are seamlessly integrated into adjacent developments and provide a host of recreational activities. Part of the PUB Singapore's National Water Agency's larger strategy is also to bring people close to the water so that they can better appreciate it. The city's blue map comprises seventeen reservoirs, 32 major rivers and more than 8,000 kilometers of canals and drains. Over 100 locations have been identified for ABC Waters projects to be implemented in phases by 2030. By 2017, over 36 projects have been completed by PUB as well as another 62 projects by other public agencies and by developers. The accompanying map shows the location of many of the projects. First initiated by the Water-bodies Design Panel led by the URA in 1989, and later formalised by PUB, the blue plan has been eventually incorporated into Singapore's Masterplan, forming an environmental overlay called the Parks and Waterbodies Plan. The Parks and Waterbodies Plan features a pervasive network of parks, open spaces, water bodies and connectors among them. With approximately two-thirds of Singapore's surface area designated as water catchment, it is important to ensure that runoff draining into its canals and reservoirs is well managed with regard to quantity and quality. On the other hand, besides water catchment and conservation *per se*. The PUB's ABC Waters Programme is a strong component of their effort to have Singaporeans take ownership and become stewards of the nation's water resources.

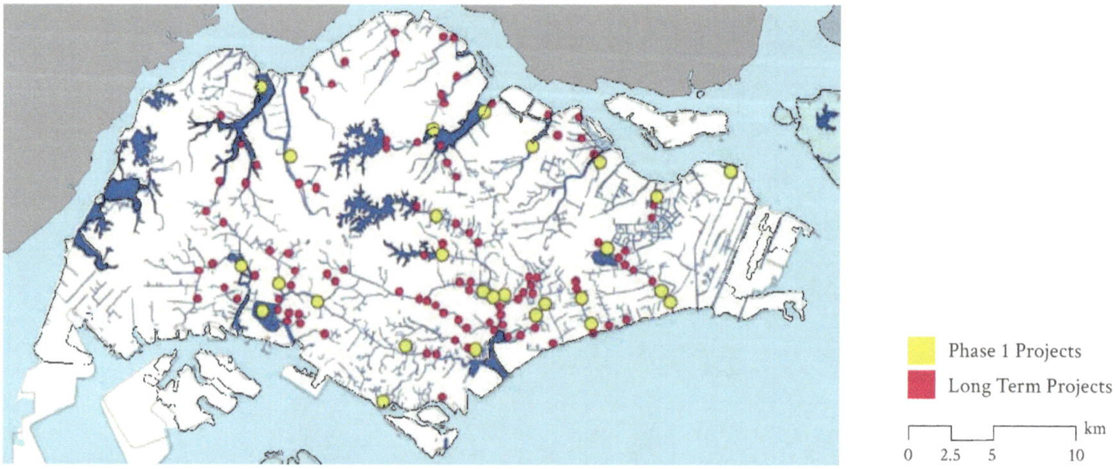

73. MAP OF THE ABC WATERS PROGRAMME PROJECTS

More technically, the PUB's ABC Waters Programme's approach to storm-water management embraces three broad conceptual domains. The first is the treatment of storm-water runoff using low-impact development approaches.[22] There, the underlying principle is to return the pre-development flow regime to an urban site and to remove storm-water pollutants by adopting ideas of catching storm-water at its origin, using simple methods, creating multi-functional landscapes and returning to natural hydrologic processes. In so doing the aim is to integrate waterways with the urban landscape and to provide a more livable and sustainable environment. This also includes accommodating flash floods adequately such as those that occurred in 2010 and 2011. In Singapore with around 2,400 mm of rainfall per annum and something like 178 rain days per year, rainfall and storm-water management is a non-trivial undertaking. Moreover, as stated throughout this book, water is also regarded as something of a precious resource. Absent this attitude Singapore would be classified as being water scarce, and ranked 140th out of 170 nations in one account of this scarcity. The second conceptual domain is the use of a source-pathway-receptor approach to storm-water management, whereby treatment starts at the origin of storm-water. The pathway component involves the traditional array of parks, canals, and drainage ditches, and receptors are comprised of flood basins and outflows where the aim is to cope with rainfall events that exceed general drainage network expectations. The third and final domain is the implementation of the ABC Waters Programme. For the ABC Waters Programme the Active (A) aspect involves new community spaces around water bodies. The Beautiful (B) part is for vibrant and aesthetically pleasing spaces, and the Clean (C) component is to improve water quality and to educate the public about the need to reduce water pollution. Drawing heavily on the Australian 'Water Sensitive Urban Design Framework', the Board's 'Engineering Procedures for ABC Waters Design Features', published in 2009 and 2011, for instance, provides extensive documentation about best management practices, local performance results and other forms of technical data. It covers, for example, at least six kinds of practices. They are: sedimentation basins, swales, bio-retention swales, bio-retention basins and rain gardens, constructed wetlands, and infiltration systems. It also deals with various aspects of such practices concerned with sizing, location, media and plant selection. Indeed, one summary account and evaluation gave high praise for the use of low-impact development practices in the ABC Waters Programme, with regard to the 'Active' and 'Beautiful' components, but less clearly with regard to the' Clean' component.

74. KALLANG RIVER AT BISHAN-ANG MO KIO PARK

Among completed projects, several stand out for their relative success or demonstration of the range of undertakings within the ABC Waters Programme. One of the largest and most ambitious was the Kallang River Project, which converted a concrete, structural storm-water conveyance into a scenic river by essentially making room for the river and use of low-impact development practices. Within this scheme of the Kallang channel, the Bishan-Ang Mo Kio Park deserves particular attention and comment.[23] Carried out by the Atelier Dreisetl, the Bishan-Ang Mo Kio Park is one of the most popular parks in the heartland of Singapore. It occupies a 62 hectare site that once incorporated concrete-lined Kallang channel, running beside the Bishan Township. Remodeling of the landscape involved converting the 2.7 kilometer long straight channel into a 3.2 kilometer long sinuous river, with natural characteristics and planting, meandering through the overall space with fluctuating water levels and ample spaces and facilities for park users. These include three playgrounds, restaurants and a new vantage point constructed from the material of the old channel. Ecological treatment of the water's edge and inclusion of appropriate planting also help to cleanse the storm-water. The return of wildlife and birdlife to the park also adds to its success. One disappointment might be found in the planting scheme which does not resonate as much as it might with tropical vegetation and general milieu. Another prominent project, also concerned with channel drainage is the Alexandra Canal, a 1.2 kilometer stretch through a dense urban area. Transformed in 2011, with CH2M Hill as the consultant, the canal now has softened banks, with a stretch decked over to create an interesting cascade and water play area.[24] Located in conjunction with the deck are a series of wetlands that engage in public learning regarding water cleansing and bio-remediation.

75. ALEXANDRA CANAL

Improvement has also been made to the venerable MacRitchie Reservoir within the Central Catchment area of Singapore. Primarily these improvements arising from the ABC Waters Programme have focused on the visitor experience and use of the reservoir. An amenity centre and structured car park have been added, along with floating pontoons for kayakers. There is also now a food and beverage outlet, prominently located on one of the hills, as well as a submerged boardwalk along which visitors can experience walking in and on water. Allowance of visitors and recreational uses within the reservoir preserve itself is a break with the past and a testament to the viability of Singapore's closed loop and treatment of its water system. A relatively new reservoir of water area is the Punggol Reservoir and particularly its Sengkang Floating Wetland installation. A primary feature at Punggol is the 'My Waterway at Punggol' that traverses the site through its town centre. The floating wetland helps to improve water quality and provides a good natural habitat for birds and fish. A boardwalk brings people close to the water so that they can observe and learn about wetland ecosystems. The wetland also provides a seamless connection between and among the clubs, sports places and parks of this new development. Some issues have emerged in the context of the ABC Waters Programme projects, critical of its operations and management. These include a certain rigidity in mindset within a comfort zone of practice and something of a fear to try even more unconventional ideas, as well as a lack of co-ordination between agencies and projects.[25] Transferability to other locales in tropical and semi-tropical regions seems feasible for several reasons. First, the current ABC Waters Guidelines are more adaptable and applicable than guidelines taken from temperate areas where low-impact development practices first started due to similarities in climate. Second, bio-retention systems and rain gardens are the most suitable low-impact features to use, that may also be adaptable and appropriate to other tropical urban areas which are hemmed in by topography or urbanization. Third, the clarity and extensiveness of low-impact development guidelines can be a boon to followers in other tropical and semi-topical settings.[26] Finally, Singapore's experience with the public in the success of the ABC Waters Programme merits careful consideration and adoption by others.

76. MACRITCHIE RESERVOIR

77. NATURE WAY ALONG PASIR RIS DRIVE 3

d.

PARK AND OTHER CONNECTORS

In 2002 Singapore launched its 'Streetscape Greenery Masterplan', emphasizing greater variety and improved scenic quality as well as identity for roadway landscapes.[27] In all there were five specific landscape treatments. They were: parkway treatments, gateway treatments, coastal treatments, forest treatments, and rural treatments. These were guided, in turn, by four design principles. They were: connectivity and convenience, luxuriousness of landscaping, landscapes enriched with elements of nature, and emphasis on quality, variety and creating a local identity from nearby circumstances or those on hand. In addition Singapore's heritage in trees and landscapes were also safeguarded as a part of the program, with heritage trees woven into the street landscapes. In fact, heritage trees became landmarked in 2001. Earlier on Singapore's 'Park Connector Network' started on the heels of the 'Park Linkage Program' of 1989. This Program aimed to use canals and open spaces of residential estates to link major parks like, for instance, the five kilometer-long Kallang River park connector. Then in 1991 the 'Park

Connector Program' was approved by the venerable Garden City Action Committee that had been in place since the late 1960s within the Ministry of National Development to ensure Garden City policies and implementation was coordinated appropriately across agencies involved. The park connector idea was also incorporated in the 1991 Concept Plan for Singapore. Materially, the park connector network optimized public open space and created value in several ways. These included: double use of road reserves; covered drains; conversion of drains into landscaped pathways; and the use of wider linear parks and green corridors and separators, like Bishan Park. Quickly park connectors became integral to creation of a 'City in a Garden' concept, communicated in, for instance, the 'Parks and Waterbodies Plan' of the Urban Development Authority and National Parks of 2002. By the end of 2009, fully 103 kilometers of park connectors had been completed, on the way to an eventual 350, or so, kilometers.[28]

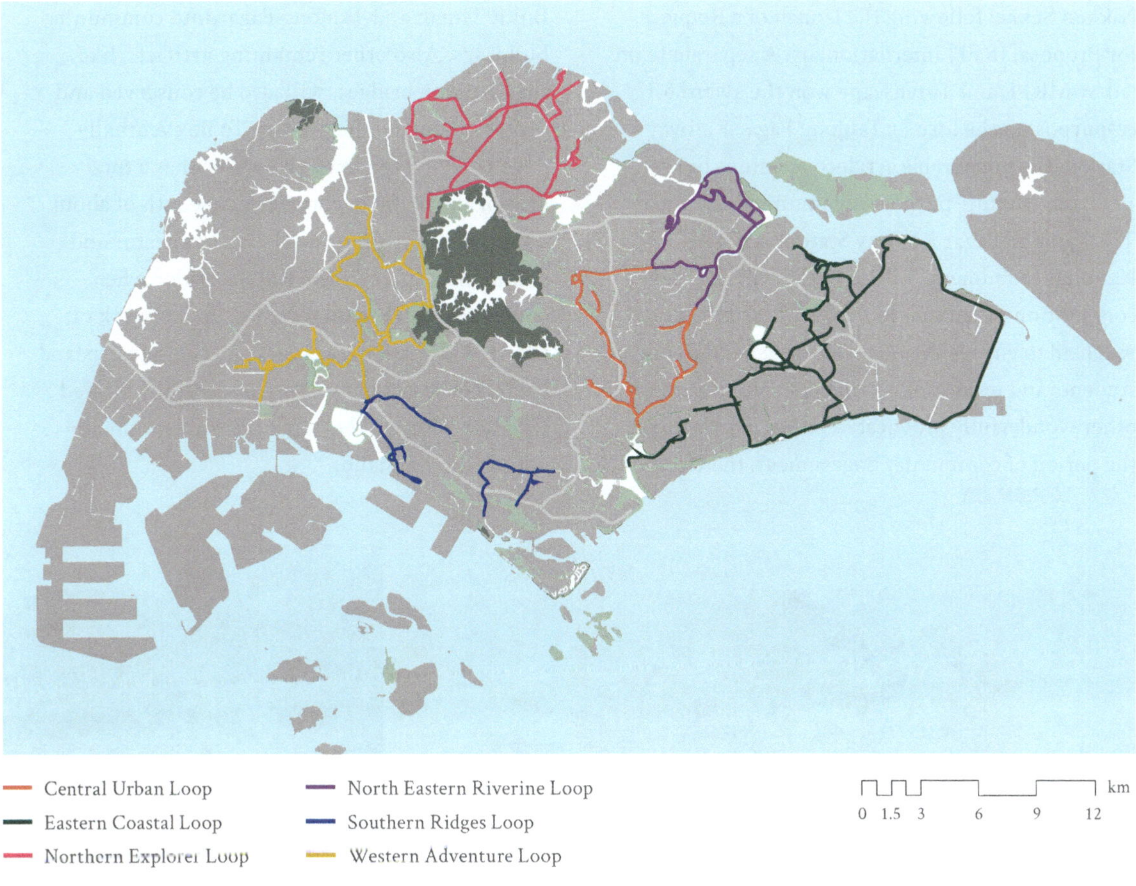

78. THE GREEN PARKWAY CORRIDOR PLAN

One of the most prominent green connector projects is the 'Rail Corridor' to be developed over the former railway line connecting Singapore to Malaysia. The railway which was operated by Keretapi Tanah Melayu (KTM) was originally constructed between 1903 and 1932 to transport rubber and tin from the Malay Peninsula for export through the port in Singapore. On 1 July 2011 the railway ceased operations and the railway lands reverted back to the Singapore Government.[29] For most of its life the railway corridor was a barrier running through the centre of Singapore Island. In re-purposing it as a green corridor it was now to be seen as a connector and much more of an inclusive and shared space, as well as a platform for community development and shared experience. In fact, within a one kilometer reach to either side of the corridor some one million Singaporean inhabitants reside, along with a plethora of community facilities, parks and heritage sites, some 58 schools as well as various places of work nearby. Recently, in 2016, a concept masterplan for the 24 kilometers Rail Corridor was awarded to a design team led by Nikken Sekkei following the launch of a Request for Proposal (RFP) internationally. A separate team led by MKPL and Turenscape won the award for re-purposing the former Tanjong Pagar Railway Station, the memorable art deco terminus built in 1932, into a multi-purpose community building. The Tanjong Pagar Railway Station is now a National Monument.[30] Prior to the RFP, an ideas competition called the "Journey of Possibilities" was held to engage members of the general public, students and professionals that resulted in many other wonderfully provocative ideas. Throughout the period of community engagement, the Urban Redevelopment Authority placed emphasis on co-visioning and co-creating the future of the Rail Corridor with the public. As far as precedents, the corridor draws inspiration from the Promenade Plantée in Paris, constructed in the 1980s and 90s, as well as the Highline in New York City as former railway lines that have been re-purposed into well-loved public spaces. History and conservation of the memory of the rail line will also be incorporated, especially with regard to the re-purposing of the railway stations at Bukit Timah and Tanjong Pagar into community buildings. Also other remaining artifacts, like the cast-iron bridges, will also be conserved and protected during the project. To be eventually developed in segments, the project has a total area of about 100 hectares over a length of about 24km, which is substantial for a connector and community space.[31] The Rail Corridor when completed will be connected to the network of island-wide park connectors and the round-island route, hence ensuring that this centrally located Rail Corridor will be easily accessible from all corners of the island.

79. THE RAILWAY CORRIDOR PROJECT

e.

THEMES AND MANAGEMENT CONSIDERATIONS

Urban greenery has recently gained popularity as an adaption and mitigation measure. Many city governments have adopted policies promoting tree-planting, preservation of urban green spaces and, more recently, green architecture. The potential benefits and services provided by greenery to the urban ecosystem include: reduction of greenhouse gas emissions, thermal comfort, improved air quality, energy-use reduction, and so on. From a social perspective greenery promotes health and a range of recreational and psychological benefits. In the case of Singapore, several themes arise with regard to the island state's greening programs. They include: biodiversity, carbon sequestration, tree modeling and management, green building installation and development, alongside sundry ecosystem studies. What follows is a summary of these activities, contributions and services, as well as an appraisal of Singapore's relative performance.

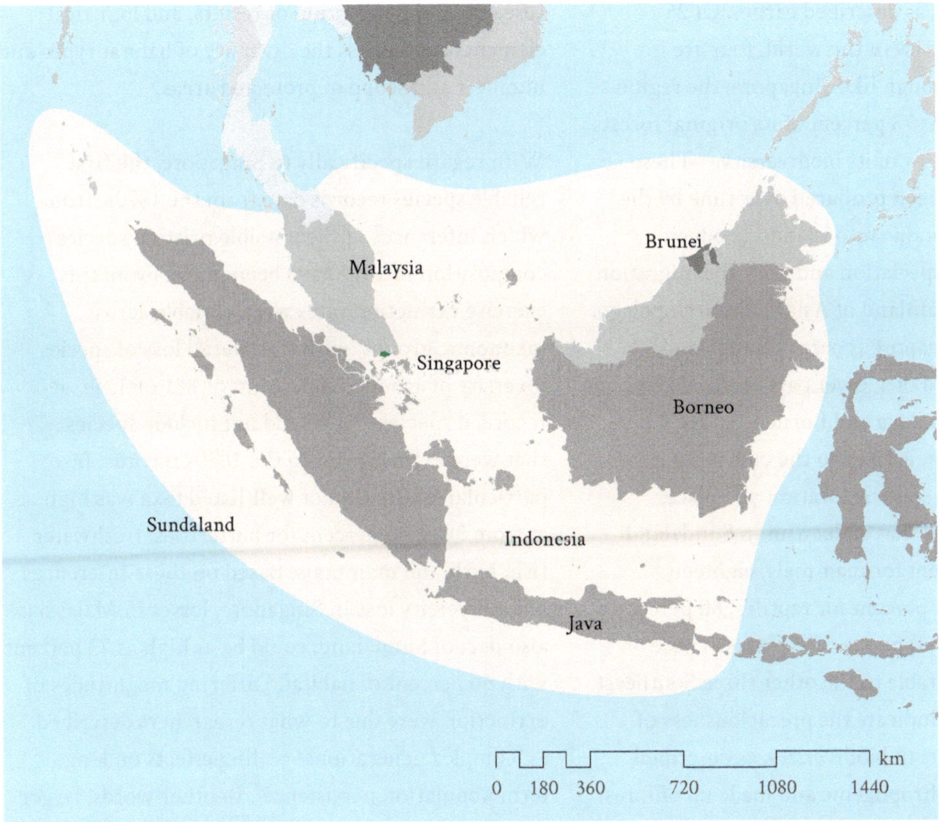

80A. SUNDALAND

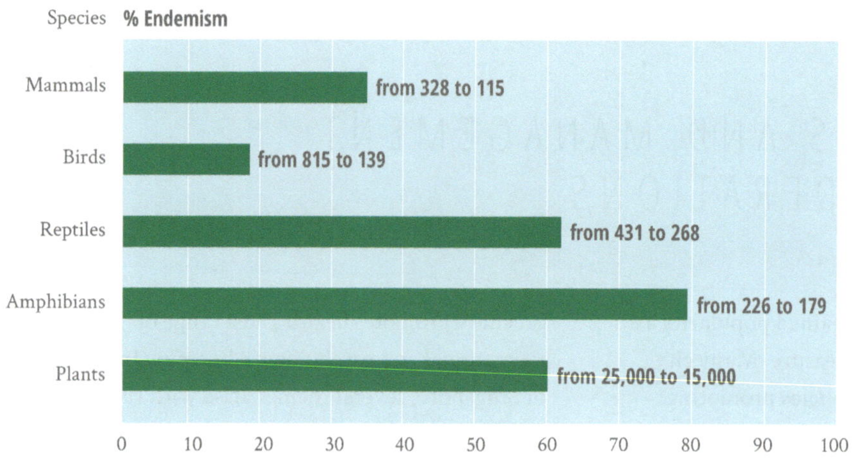

80B. ENDEMISM WITHIN SUNDALAND

Tropical ecosystems are exceptionally rich and exclusive reservoirs of much of the earth's biodiversity and Singapore is no exception, except for the historically high rate of deforestation that occurred from 1819, as described earlier. Of 25 hotspots of biodiversity in the world, four are in Southeast Asia, although like Singapore, the region could lose as much as 75 percent of its original forests by 2100 and 42 percent of its biodiversity.[32] These four hotspots have been produced over time by the conversion of mountains into islands, producing ideal condition for speciation and with the migration of biota from the mainland of Asia to the archipelago. Of the hotspots Singapore is part of Sundaland, otherwise comprised of a lower part of the Malaysian Peninsular, Sumatra, Java and Borneo. As elsewhere, the rate of extinction is fixed to the endism of the hotspot or to its specific location and place. Recent study showed that endemism in Sundaland was around 35 percent for mammals, eighteen percent for birds, 61 percent for reptiles, 80 percent for amphibians and 60 percent for plants. These numbers are comparable to the other three Southeast Asian hotspots and indicate the precariousness of the resource. Threats to biodiversity, as described here, are largely anthropogenic and made up of forest conversion, forest fires, hunting for bush meat and wildlife trade. Conservation challenges include: social, scientific and logistical elements. Social aspects are population growth, poverty, shortages of conservation resources and corruption.[33] Scientific elements are neglect of research, low levels of research and publication of results, and logistical elements encompass the diversity of habitat types and numbers and scope of protected areas.

With regard specifically to Singapore, the first reliable species records date from the 1870s, from which inferences of the possible pristine species composition in 1819 have been made. From this exercise extinction rates were variable across taxonomic groups, with the overall loss of species diversity of around 28 percent, or 881 of 3,196 of recorded species.[34] This did not include species that were extinct prior to the 1870s records. In particular, extinction of well-listed taxa was high, at from 34 to 43 percent for butterflies, freshwater fish, birds and mammals. Based on these inferences of biodiversity loss in Singapore, losses in Malaysia, also part of Sundaland, could be as high as 73 percent with 60 percent of habitat. Differing magnitudes of extinction were due to what researchers described as "complex generational scaling effects on long-term population persistence". In other words, larger organisms have more habitat to support viable populations, but live longer. Most extinctions appear to have occurred among biota related to forest habitats, with 33 percent rates of extinction, compared to seven percent for those species that preferred or tolerated open-forest edge habitats. The predominant cause of Singapore's extinctions were rapid and large-scale habitat destruction initially, followed by urban development. Habitat loss, fragmentation and modification caused extinction by wrecking breeding and feeding sites, increasing predation, soil erosion and thicket loss, as well as dispersal exclusionary edge effects. However, hunting and collecting was also involved. Large vertebrates, like the Malaysian Tiger, were hunted as menaces to human life and livestock. As noted earlier the last tiger was shot in 1930, of a total of some 99 animals, with 125 losses of human life. Also the shelling of natural preserves during World War II probably had a detrimental effect on forest fauna. In fact, future prospects look bleak for Singapore's surviving biodiversity, according to some experts. Fully 77 percent of the island's species continue to be threatened, based on World Conservation Union regional listing criteria. In short, all existing pools and resources must be protected.

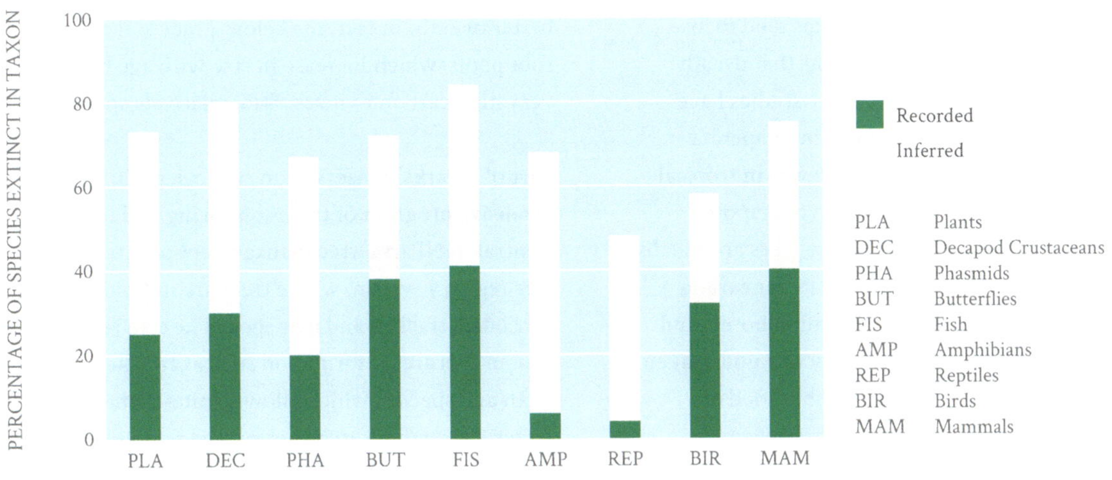

81. SPECIES EXTINCTION BY TAXONOMIC GROUPS

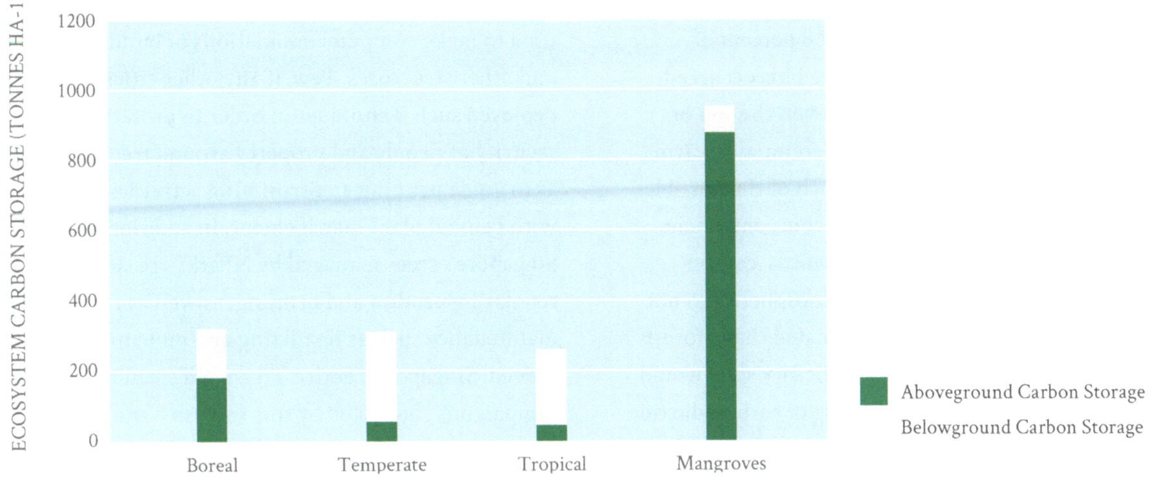

82. CARBON SEQUESTRATION BY FOREST TYPE

To date there is little empirical evidence showing the effectiveness of urban vegetation to reduce greenhouse gas emissions or concentrations of airborne pollutants.[35] This includes evidence demonstrating the direct removal of carbon dioxide from the atmosphere by urban vegetation. There are a number of reasons for this, but a complete assessment of the contribution to carbon sequestration by urban greenery needs to consider both the carbon accumulated by trees and by the soil respiration. In other words, contributions from above ground vegetation and underground soil processes. Many studies show, for instance, that recently disturbed ecosystems tend to lose carbon, unlike old-growth forests that usually act as carbon sinks. To date more studies have been made of these phenomena in temperate climatic circumstances, but far fewer in tropical and semi-tropical climes, such as Singapore's tropical rainforest climate. There trees are usually evergreen and, therefore, potential for carbon dioxide assimilation is larger than in boreal and temperate forests. A recent study was undertaken, including the Telok Kurau neighborhood of Singapore. There the major anthropogenic contribution of carbon dioxide came from vehicular traffic, followed by human metabolic respiration, especially given population pressures and density. Data from tests suggested that above ground vegetation sequestered 7.8 percent of the total emitted carbon dioxide. This occurred mainly by photosynthesis. However, the soil or below-ground efflux from the perennially warm and humid soil cancelled out much of the possible carbon uptake, making the biogenic component a net emission source. In other words, carbon dioxide was recycled into the atmosphere and not sequestered. It was further estimated that a forest area 30 to 50 times the area of the city state would be needed to offset the 38,790 Cg of carbon dioxide equivalent, or 97 percent of emissions. Over long time scales carbon storage will depend on the amount of urban expansion, greenery management and carbon allocation to biomass and organic material. Each year biomass is transferred to the soil and removed from the urban ecosystem through pruning and debris collection. In Singapore, although large trees account for 36.8 percent of all trees they contain 95.3 percent of the biomass and, therefore, carbon. When taking vegetation and soil together the biogenic component was found to add 4.4 percent extra carbon dioxide in the Telok Korau neighborhood of Singapore. Mangrove forests, however, also to be found in Singapore perform far better overall. In fact, the below-grade soil and dead root pools, which increase in size with age have a very high carbon to mass ratio, relatively speaking.

Part of NParks' conservation strategies is an extensive program of tree monitoring and research. Central to NParks' tree management program is a tree registry system, where trees are individually geocoded, tracked and managed. The tree registry system captures information such as tree height, girth and species, which allows finite-element structural analyses to be performed remotely to determine the stability of the trees. The results of the analyses are used to guide pruning decisions to strengthen the resilience of the trees against strong winds.[36] Finite-element models are more usually used to make computer simulations of buildings and other structures. Few, if any other cities, have deployed such techniques in order to protect the security of people and property around trees, as well as to guide pruning and trimming activities along with general plant management. In addition, all Singapore's trees managed by NParks are subject to regular inspection and pruning, as well as routine maintenance such as fertilizing and mulching. Indeed, Singapore's entire 'green and clean' campaign is supported by this intensity and style of monitoring, data keeping and research.

83. GREEN WALLS FROM THE SINGAPORE INSTITUTE OF TECHNICAL EDUCATION

The greening of roofs and walls as 'living' roofs and walls has become a relatively prominent feature of Singapore's urban landscape. In fact interest in such 'living' wall and roof installations dates back to the 1960s and 70s. Technically, there are several ways in which green living walls can be provided on buildings and other structures, such as bridges and flyovers.[37] Most common are wall mounted systems, where panels or similar components comprised of vegetation are attached to the outside of walls or structures. There are also free-standing systems, using trellises or frameworks that allow plants to grow up building facades. Then there are also impregnated systems of bio-soil, or similar confections, incorporated with conventional wall materials, like impregnated concrete. Wall-mounted systems and free-standing systems allow sufficient depth for deeply-rooted plants and also well-irrigated systems of plants growing up from ground level. The benefits of green living wall and roof systems are numerous. These include: reduced heating needs by adding mass and thermal resistance and also the reverse with reduced cooling. Reducing or holding storm-water runoff can also occur, alongside of natural habitat creation and a certain capacity to filter air pollutants. Temperature fluctuation reductions can be as much as from 10 to 60 degrees centigrade to from 5 to 30 degrees centigrade.

As of 2017, Singapore had 100 hectares of vegetated roof space, with concerted policy efforts by the NParks and the URA to promote and advance green living walls and roofs in 2009 and 2011 respectively.[38] Introduction of the 'Landscape Replacement Policy' meant that new developments in certain areas were and are required to provide landscaped areas to make up for all greenery and landscape loss on the ground. Consequently a landscaped area equivalent to the area of a development site must be provided for in the form of landscaping at ground level, on sky terraces, mid-levels and roof-top gardens. In fact sky terraces have been encouraged in larger and loftier forms and as communal spaces. Through its 'Skyrise Greenery

Incentive Scheme', NParks agreed to cover 50 percent of installation costs. Prominent among living green buildings in Singapore are two hotels by the local architecture firm of WOHA.[39] One is the Parkroyal on Pickering, on the edge of Chinatown, replete with green roofs, wall elements and façade trellises. The other is the Oasia Hotel Downtown, which is fashioned as a high-rise tower in the form of a trellis wrapping of the overall building shell. Elegantly shaped this red-painted trellis hosts 21 species of flowering plants, greening up the side elevations towards a partially-enclosed sky terrace with a large mid-level opening and mid-level terrace within the overall volume of the building. In recent years many other buildings have incorporated similar living wall and roof systems. In addition, considerable effort has been given to improve the overall resource demand and residual production efficiencies of buildings, commonly covered in LEED or similar rating systems.

84. PARKROYAL ON PICKERING HOTEL BY WOHA

85. OASIA HOTEL BY WOHA

f.

TOWARDS A BIOPHILIC OUTCOME

At a certain basic level there are two types of green landscapes in Singapore. The first belongs to its natural systems, catchments, native forests and so on. The second are artificial constructs of parks, gardens, streetscapes, canals and skyline terraces. From its inception and particularly during the trend from Singapore as a 'Garden City' to a 'City in a Garden' and on to a 'City in Nature', NParks, the authoritative oversight and driver, has been restoring ecosystems into artificial constructs and ecological processes with regard to the natural systems in place. The overall aim is to have Singapore functioning as a fully urban ecological system. Further, part of the aim is to connect and re-connect people with nature and to bring a more profound sense of being with nature into their lives. In so doing NParks is highly proactive, constantly searching for and identifying elements of the built environment that can be associated with greenery or all it portends. In fact, Singapore is unique among nations in the way it manages greenery. This derives from the scope of activities, the variety of activities, and the technical sophistication and scientific, botanical and other research backing up its greening efforts. The preceding sections have described these efforts and the almost inexorable trend and even obsession that has gripped the island state in these regards and particularly over the last half century.

Lurking closely behind Singapore's efforts to connect its citizens with nature is that such association is beneficial and in fundamental ways. In a word, it is about becoming biophilic or certainly verging towards that condition. Simply put, biophilia suggests that humans possess an innate tendency to seek connections with nature and other forms of life. It is a state of being, popularized by luminaries like Edward O. Wilson in his book with a similar name of 1986.[40] More succinctly he describes biophilia "as the urge to affiliate with other forms of life". From this it closely follows that 'biophilic space' strengthens and supports social psychological capacities of life, spanning acts that unburden cognitive systems and further sensorial experience. More prosaically, biophilic undertakings, by creating strong connections between nature and man-made environments has benefits. To date there is evidence to suggest that this can include helping to make office workers more productive, encouraging children to learn and develop, as well as helping hospital patients to get better. There it is very much about natural light, views of nature, presences of plant life and the use of natural materials, textures and patterns.

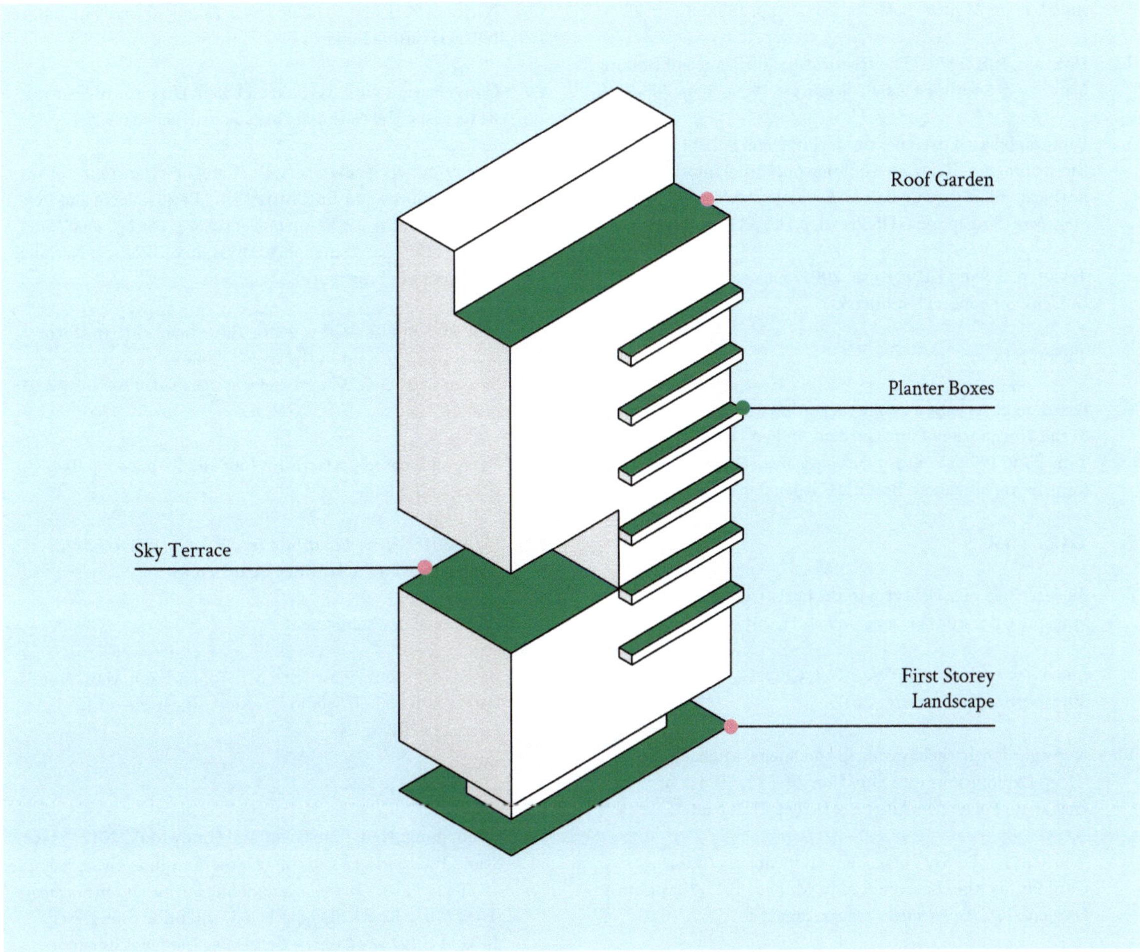

86. DIAGRAM OF THE LANDSCAPE REPLACEMENT POLICY

endnotes

1. Soderstorm, M. 2001. *Recreating Eden: A Natural History of Botanical Gardens* (London: Vehicule Press) and Hill, Arthur W. 1915. "The History and Functions of Botanic Gardens". *Annals of the Missouri Botanical Garden*,pp. 185-240.

2. Holttum, R.E. 1970. "The Historical Significance of Botanic Gardens in Southeast Asia". *Taxon*, vol 19, no 5, pp.707-714.

3. Paragraph based partially on Taylor, Nigel P. 2014 "The Environmental Relevance of Singapore" in Timothy P. Barnard. Ed. *Nature Contained: Environmental Histories of Singapore* (Singapore: NUS Press), p.117, p.118 and p.119.

4. Taylor, p.123 and Kiew, Ruth. 2001, *Singapore Botanic Gardens* (Singapore: Landmark).

5. Taylor, p.18.

6. Based on observation and presentation during a field trip to the Singapore Botanic Gardens on Feb 11, 2017. Also, see Yam, T.W. 1995.*Orchids of the Singaporean Botanic Gardens* (Singapore: Singapore Botanic Gardens).

7. Taylor, p.130.

8. As described on a field trip to the herbarium of the Singapore Botanic Gardens on Feb 11, 2017

9. Interview with Dr. Tan Wee Kiat, CEO, Gardens by the Bay, Singapore on 15 February 2017.

10. www.gardensbythebay.com.sg and Spens, Michael. 2007. "Deep Explorations into Site/Non-site: The Work of Gustafson Porter." *Architectural Design*, 77:2, p.66-75.

11. "Gardens by the Bay", http://en.wikipedia.org/wiki/Gardens_by_the_Bay and Reddy, Mohan. 2013. "Bay South Garden". *ICC Plan Review Service*, August 18.

12. Tan Wee Kiat, interview, 15 February 2017.

13. "Plans for the next century". *Straits Times*. April 27, 1987, p.2 and Justin Zhuang. 2016. "Marina Bay". *Singapore Infopedia*. (Singapore: National Library Board).

14. Paragraph based in part on Barnard, T.P., 2014

15. National Parks. 2017. *Parks and Nature Reserves* (Singapore: NParks).

16. Based on a field trip to the Eco-Link on Feb 10, 2017

17. As observed on a field trip to the Eco-Link and Bukit Timah Reserve.

18. National Parks.2017. *Bukit Timah Nature Reserve* (Singapore: NParks) online bulletin.

19. Conversation with Ms Sharon Chan, Director of Central Reserves at the National Parks Board, February 2017.

20. https://en.wikipedia.org/wiki/Category:Nature_reserves_in_singapore and Lim, Shirley S.L.; Peter k. l. Ng and Leo W. H. Tan et al. 1994. *Rhythm of the Sea: The Life and Times of Labrador Beach* (Singapore: Division of Biology: National University of Singapore).

21. Lim, H. S. and X. X. Lu. 2016. "Sustainable urban storm water management in the tropics: an evaluation of Singapore's ABC Water Program". *Journal of Hydrology*, 538, p. 842.

22. Paragraph based partially on Lim and Lu p. 843, p.844, p.845, and p.846.

23. PUB. 2014. *Active Beautiful Clean Waters Design Guidelines* (Singapore: Public Utilities Board), p.88

24. PUB, n.d. *Alexandra Canal.*

25. Based on an interview with Mr. Tobias Baur, Managing Director, Ramboll Studio Dreiseitl Singapore on 10 February 2017.

26. Lim and Lu, p. 859.

27. Boon Siong, Neo; June Gwee and Cindy Mak. 2012. "Case Study 1: Growing a City in a Garden" in June Gwee. ed. *Case Studies in Public Governance: Building Institutions in Singapore* (New York: Routledge), pp.11-63, and Boon Siong, Neo. 2009. *A City in a Garden – Developing Gardens* (Singapore: Centre for Governance and Leadership, Civil Service College), p.25.

28. Boon Siong, p.21.

29. Based on an interview with Mr. Tan See Nin, Senior Director, URA, on 16 August, 2017.

30. Tan, power point presentation.

31. Tan, power point presentation.

32. Sodhi, Navjot S.; Lian Koh; Barry W. Brook and Peter K.

33. L. Ng. 2004. "Southeast Asian Biodiversity: an impending disaster. *Trends in Biology and Evolution*, vol 19, no 12, p.645.

Sodhi, p.568.

34. Paragraph based partially on Brook, Barry W.; Navjot S. Sodhi and Peter K. L. Ng.2003. "Catastrophic extinctions follow deforestation in Singapore" *Nature*, vol 424, p.420 and p. 422.

35. Paragraph based partially on Velasco, Erik; Matthais Roth; Leslie Norford and Luisa T. Molina. 2016. "Does vegetation enhance carbon sequestration? *Landscape and Urban Planning*, vol 148, p.99, p.110, p.111 and p.105.

36. Based on an interview with Mr. Kenneth Er, CEO, NParks and Dr. Leong Chee Chiew, Deputy CEO, NEA on 10 February 2017.

37. For example, Greenliving Technologies International. 2017. *Green Living Roofs, Green Living Walls, Vertical Farms*, (Rochester, New York: Newsletter).

38. Er and Chiew, interview.

39. Busenkell, Michaela and Cachola Schmal, Peter. 2012. *WOHA: Breathing Architecture* (Munich: Prestel).

40. Edward O. Wilson. 1986, *Biophilia* (Cambridge, Mass: Harvard University Press).

chapter

06

WAYS FORWARD

P. G. Rowe and L. Hee, *A City in Blue and Green*,
https://doi.org/10.1007/978-981-13-9597-0_6

Beside Singapore's recent glamorous rise in notable architecture and urban design, centres of activity, event curation and other urban enterprises, the rise of the 'green-blue' planning and implementation since the early 1990s is perhaps its most truly distinctive and notable achievement in comparative terms with elsewhere in the world. By all accounts and certainly those discussed here, it has been and is truly remarkable, making Singapore a world leader. Much of this and further success can be attributed to the particular nurturing of Singaporean habits of mind, or what can be referred to as '*Kiasu*' ("afraid of losing" and "afraid of being static and needing to move on") with large doses of pragmatism, incremental effectiveness, doggedness and collective independence of opinion. Lurking behind is also a strong belief in the perfectibility of cities and that well-laid plans can be successfully carried out along with the technology needed to support them. While often leading to clean-cut and relatively narrow norms and ways of life, overall betterment has ensued for many if not most. Leadership, political will, visionary insight, clarity and directness in co-ordination among public and private agencies has come to the fore, with little wasted energy and outcomes. In the future, both internal and external challenges will undoubtedly emerge. Further buy-in by the public to Singapore as a truly 'city in nature' will require attention and public conversation. The result, however, also seems likely to be one of a kind, or if the rest of the world is shrewder in its choices, the first among many of its kind.

More squarely within the frame of reference concerning 'blue-green' planning and performance, Singapore appears almost certain to achieve many of its objectives. To be sure by 2061 if not earlier, some semblance of domestic water sustainability will have been accomplished, though not in the virtual realm beyond the island. Consequently, Singapore will rightfully take its place as a leader in the world of the water sector. Its domestic water sector initiatives will have continued to penetrate world markets and to set new standards of accomplishment elsewhere. The closed-loop, used water scheme of Singapore will be copied and the island state's expertise drawn upon and widely acknowledged. The 'green' component will also not go unnoticed, particularly as Singaporeans make their way into a more biophilic state with nature and begin to pioneer and make attractive the liberating effects of dense living and working within intense combinations of tropical flora and fauna. The necessary higher degree of acquaintance of Singaporeans with nature in order to secure their safe future in water will pay off and become more than an alibi for 'blue' with 'green' and a distinctive notable feature of the island state's urban landscape.[1] The integration required will put everyone more or less on the same page, so to speak, helping to further forge the degrees of inter-agency agreement and collaboration that have already been noteworthy and distinctive, but also in other offices as well. However the moral high-ground that Singapore seems likely to achieve among nations in pursuit of economic and environmental sustainability seems more likely to be achieved than not, pushing the young Republic further into the forefront.

87. THE GREEN BLUE PLAN ARISING FROM THE 1991 CONCEPT PLAN

a.

SUCCESSFUL INGREDIENTS

The ingredients for Singapore's success in these regards are several and interconnected. First, Singapore's governance is characterized by strong political will, visionary leadership, clear policy directions, whole-of-government collaboration, public-private partnerships, notable institutional capacities and what some have termed an ideology of pragmatism.[2] The extraordinary vision of Lee Kuan Yew from the early days of the Republic have already been commented upon and were pivotal to Singapore's rise and success along with the political will and clear insight that was also required, The success of single political party rule was constantly earned and not docilely granted, as witnessed in the most recent general elections in favor of the People's Action Party (PAP).[3] At a time when there was a certain amount of uncertainty about futures, votes were resoundingly cast for those who had made a difference and seemed to be well-known as distinct from unknown quantities in the mind's eye of the electorate. Such is to be expected rather than to be seen as unusual or somehow extraordinary by way of an outcome.

88. FROM A 'GARDEN CITY' TO A 'CITY IN NATURE'

Interagency collaboration was also necessary in order to capitalize on vision and leadership. Further, this occurred among several significant programmatic aspects of Singapore's 'green-blue' planning and implementation and for a number of reasons. In the broad scheme of things this flowed from the persistence, discipline and clarity of the Concept Plans and Masterplans, beginning in 1971. It was also strongly manifested in programs promulgated from lead agencies, like the PUB in the ABC Waters Programme, of NParks Streetscape Greenery Masterplans in conjunction with the URA. A sense of mutual respect, shared urgency of vision and technical capability underpinned these ventures. Throughout there was a high reliance on scientific and technical knowledge combined with real pragmatic reckoning with need. Arguments were thus reduced more to substance than to form, making collaborations easier. One sector's interests portended leadership, perhaps, but not a totalizing or dominant role.[4] This also enabled appreciably the whole-of-government approach to broad issues of interest and concern. A capacity for specific purpose agencies to morph, widen or focus their interests also played a role in the continuation of strong collaboration, especially as trends in both the water and green space sectors began to shift. For example, when the earlier version of the National Parks Board was merged with the Parks and Recreation Department in 1996, and became NParks today is a case in point and when the PUB foregrounded the ABC Waters Programme so strongly and forthrightly.

Public awareness and campaigns to keep the public interested and engaged also contributed to Singapore's success. The public at large were constantly made aware of environmental concerns and responsibilities from the early 'clean and green' movements to later ABC Waters Program. These were also metered out in bite-sized pieces, as it were, and for different constituencies on the way to reaching everyone. Again a clear case in point is when the ABC Waters Program's engagement began with children and then moved on to the adult populations of the parents. The campaigns were also attractive by way of presentation and purpose. The tone of urgency by not being either too lax or too strident was also important. Further, the tracking of specific campaigns in keeping with the temporal roll-out of broader programs also helped to engage the public through its relevancy, adding to the comfortable bite-sized messages being communicated.

Then too, something referred to as *kiasu* as a habit of mind of Singaporeans in general could be seen to lurk behind all these other ingredients, keeping the whole community, or most of it, synchronous or on the same page as it were. *Kiasu* is roughly translated as "afraid to lose", based on the Hokkein *kia* meaning 'afraid' and *su* meaning 'lose'.[5] As such it can imply possession of a grasping and selfish attitude. But moving further into the etymology of Singlish it can also mean an 'over-cautiousness' and 'fear of failing', as well as more positive senses of being *kiasu* in order to achieve something and to get ahead. As a general state of being kiasuism is perhaps closest to the American term 'paranoia' or an attitude driven by fear.[6] It can also convey the idea of risking as little as possible, which brings it close to a kind of pragmatism. Commonly used in Singapore often with an intended negative connotation, in some contexts it conveys a certain stalwart resolve to move ahead. One such context was during the early uncertain days of the Republic and a time of scarce resources. Without over-construing collective consciousness in this direction, *kiasu* combined with strong doses of pragmatism are evident in the almost dogged, incremental and technological pursuit of the ends of water security and the strong almost biophilic move towards tropical nature. Pragmatism is, after all, an outlook and approach that assesses the truth of theories and beliefs about situations primarily in terms of practical application. As one noted official commented, there was nothing fluky or overly chance laden in the Singaporean approach to environmental management, water security and many other issues. Strong leadership, interagency collaboration, public awareness, all with a dash of *kiasu* and practical reckoning with reality, is what made the difference in Singapore.

b. PUBLIC PARTICIPATION

What might be termed state-aided participation on the part of the general public was and is visible in Singapore's planning processes. Moreover, in its disciplined practice it is different from other forms that often appear bottom-up, spontaneous and unruly. Specifically in the development towards a 'Garden City,' a 'City of Gardens and Water,' the Singapore government has actively attempted to instill in their residents a sense of environmental consciousness, communitarian values and social responsibility over greening activities.[7] This public engagement takes two forms: raising public awareness, as described earlier, and public participation and consultation through, for instance, focus groups or workshops.[8] Campaigns that aim to inform the Singaporean public and raise awareness generally have an additional goal of creating ownership over a specific project or government policy. In fact, campaigns will often precede the introduction of an environmental or public health law.[9] Examples of campaigns, as noted earlier, include Singapore's long-running 'Clean and Green' Week, the recently launched Park Connector Appreciation Day, and the Annual Tree Planting Day. The Community in Bloom Program, in which Singapore's NParks aims to promote a gardening culture among residents is another example that contributes to the overall greening strategy and has the added explicit objective of encouraging social cohesion, even if the extent to which the later has proven successful is contested.[10] Also as noted earlier a second major public outreach method is through the educational system. By educating schoolchildren, the wider public is informed.[11]

Public participation in green and blue projects in Singapore takes place through state-initiated focus groups or, for instance, design workshops at a project level. For example in the development of the Singapore Green Plan of 2012, setting out Singapore's approach to environmental sustainability for a ten year period, focus groups formed an integral part of the participatory process.[12] In addition there have been some examples of proactive interventions by Singapore's civil society. Although somewhat unconsolidated, Singapore's civil society influenced the determination of Sungei Buloh Freshwater Swamp as a bird sanctuary and natural reserve, as well as the reversal of plans to reclaim the Chek Jawa Wetlands on Pulau Ubin Island. As far as on-going projects are concerned, the Rail Corridor coordinated by the Ministry of National Development and the Urban Redevelopment Authority, is seen as a unique opportunity to engage the public, with the Nature Society instigating strong public engagement resulting in the consolidation of a Consultation Group comprising nature, heritage and other interest groups. As described briefly earlier, public engagement in this project included ideas competitions, design workshops, and an online portal. Another key example is the ABC Waters Program, in which public engagement played a crucial role in ensuring community buy-in and ownership over the water and park assets after implementation.[13]

One contention in relation to public participation in Singapore centres on the lack of participation from the working and middle classes, or really, the majority of the Singaporean population. As described by one commentator, public influence on planning processes in Singapore is contingent on three factors. Citizens have to abide by conditions and boundaries set up by the state.[14] This is particularly relevant to issues of ethnicity and religion. Then also participation should take place from a constructive standpoint with a purpose of seeking consensus rather than contention and, finally, citizens should accept that their views may be scrutinized and challenged. As stated by some other authors, this may cause less informed or skilled citizens to be discouraged from actively participating in planning and consultation processes, leading to a skewed representation of the general population with a predominance of so-called 'super-citizens'. This is a term coined to describe those citizens that are well-informed, adhere to the requirements of engagement, and are likely to be professionally related to the field of concern, such as practitioners, investors and academics.

C. FUTURE CHALLENGES

In the context of Singapore's 'blue and green' activities, the challenges likely to be faced in the future seem to emanate primarily from beyond its borders, although certainly engaging internally with the capsule ecology the island estate has developed for itself. Among the trans-boundary conditions affecting Singapore there are at least three particular phenomena. They are: climate change and its various manifestations; externally posed threats to public health and environmental quality; and the risks posed by civilian nuclear proliferation within the region. Among these, the effects of climate change probably looms largest, although the other two cannot be so easily dismissed. Geographically, Singapore is surrounded and in close proximity to Malaysia, from which it broke away and Indonesia, close by on the east and southwest. Further afield are other Southeast Asian nations, broadly comprising the ASEAN group of member states.

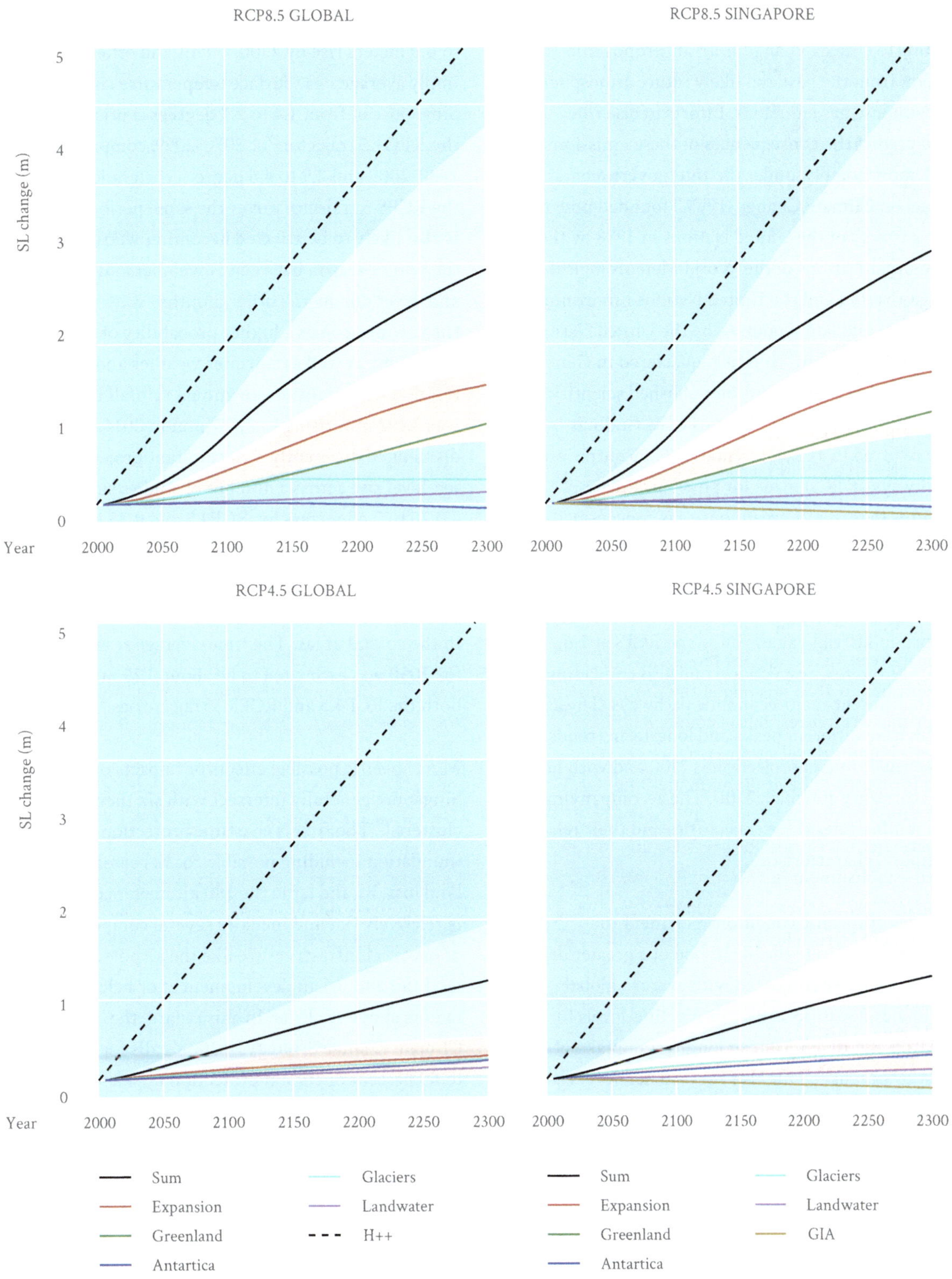

89. GLOBAL AND SINGAPORE CLIMATE CHANGE SCENARIOS (RCP8.5 AND RCP4.5)

An overwhelming amount of scientific evidence points to climate change as an anthropogenic event, from the past and likely future atmospheric greenhouse gas emissions. Efforts to describe and predict the consequences of these emissions fall, most notably, under the Intergovernmental Panel on Climate Change (IPCC) founded under the auspices of the United Nations in 1988 by the two organizations of the World Meteorological Organization and the United Nations Environment Program and later endorsed by the United Nations General Assembly.[15] It is headquartered in Geneva and bases its assessments on published scientific literature. Various distillations of its findings are referred to as 'Representative Concentration Pathways' (RPCs) made for possible climate futures in terms of greenhouse gas concentrations expressed as carbon dioxide equivalent units.[16] Generally, there are four pathways with ranges of radiative forcing values into the future and relative to pre-industrial values. They are: RCPs at 2.6, 4.5, 6.0 and 8.5, respectively and also coinciding with different socio-economic pathways. These, in turn, have different peaks and long-term trends. The usual time of projection is 2100 and with far less certainty attached, 2300. The accompanying graph illustrates these trajectories and their relative temporal characteristics.

Studies of specific conditions pertinent to Singapore indicate the likelihood of a greater degree of sea-level rise compared with average global estimates due to so-called 'in-pipeline' levels for RCP4.5 and RCP8.5 due to long-term uptake of heat by oceans.[17] This also suggests sea-level rises in Singapore slightly higher than global averages by about five percent. Modeling suggests a lower bound of sea-level rise in 2300 under RCP4.5 at 0.36 to 2.10 meters and an upper bound under RCP8.5 of from 0.94 to 5.48 meters. Applying sensitivity testing to these results for adaptive purposes, such as construction of seawalls, an upper limit range of 1.0 to 2.0 meters rise for the 21st, 22nd, and 23rd centuries is suggested, with from 3.0 to 6.0 meters rise by 2300.[18] This and other studies imply average near surface temperature rises in Singapore of from 1.4 to 2.7 degrees centigrade under the RPC4.5 trajectory at 2070-2099, compared to 1980-2009 and 2.9 to 4.6 degree centigrade rises for the RCP8.5 trajectory over the same period. Rainfall is also likely to be affected becoming wetter during the winter season between November and January and dryer during the other months. With these fluctuations comes a higher probability of drought conditions as well as stormier weather and flash floods. Statistically, mean annual rainfall in 1980 was 2,192 mm rising to 2,727 mm in 2014, though distributed differently. Concomitant projection of sea-level rise across a number of studies suggest a 0.25 to 0.65 meters for RCP4.5 and 0.35 to 0.76 meters for RPC8.5 by 2100. Most of Singapore Island is 15 meters or more above sea level, with about 30 percent less than five meters above sea level mainly in the coastal areas. The time-average sea-level rise for 2050 was estimated to be about 0.25 meters for both the RCP4.5 and RCP8.5 trajectories.[19]

More specific possible effects or impacts on Singapore generally intersect with six thematic clusters.[20] The first is coastline protection from inundation including possible coastal erosion and land loss. By and large the 180 kilometers of coastline is relatively flat and includes several components of essential infrastructure like the airport, port facilities and urban development at or below two meters above sea level. In this regard, the 'National Climate Change Strategy of 2008' calls for 70 to 80 percent of coastal protection by sea walls; although with inundation and potential loss of mangroves.[21] A second thematic domain is water resource management comprising flooding and deterioration of the water supply. Extreme weather events will likely cause inland flooding, especially during the monsoon season. Already the PUB has made significant reductions in flood-prone areas, from 3178 hectares in the 1970s down to 98 hectares in 2010 and to 40 hectares by 2013. Adverse effects on reservoirs may come by way of higher levels of evaporation plus possible contamination through temperature rise, inducement of algae blooms and saline intrusion. However, in the latter case variable salinity level treatment processes already in place will be effective. The third thematic cluster concerns changes in biodiversity. At around 1.5 to 2.5 degree centigrade increases in temperature there will be impacts on plants and animals, as well as on soil formation, nutrient storage and pollution absorption. Public health, the fourth realm, will likely be affected through vector-borne disease susceptibilities, like dengue fever, and through increased heat stress, especially on elderly and infirmed inhabitants. Adverse impacts on the fifth cluster concerning buildings and related facilities, as well as the sixth cluster concerned with network infrastructure, will come largely by way of inundation within lower-lying areas.[22] To these six areas could also be added potential heat-island effects as more urbanization occurs, with concomitant impacts on public health from heat stress and on Singapore's energy budget for cooling. Also more indirectly climate change will pose an outside potential threat to Singapore due to fluctuations in supplies and prices in the global food supply. The island state currently imports as much as 90 percent or more of all its food.

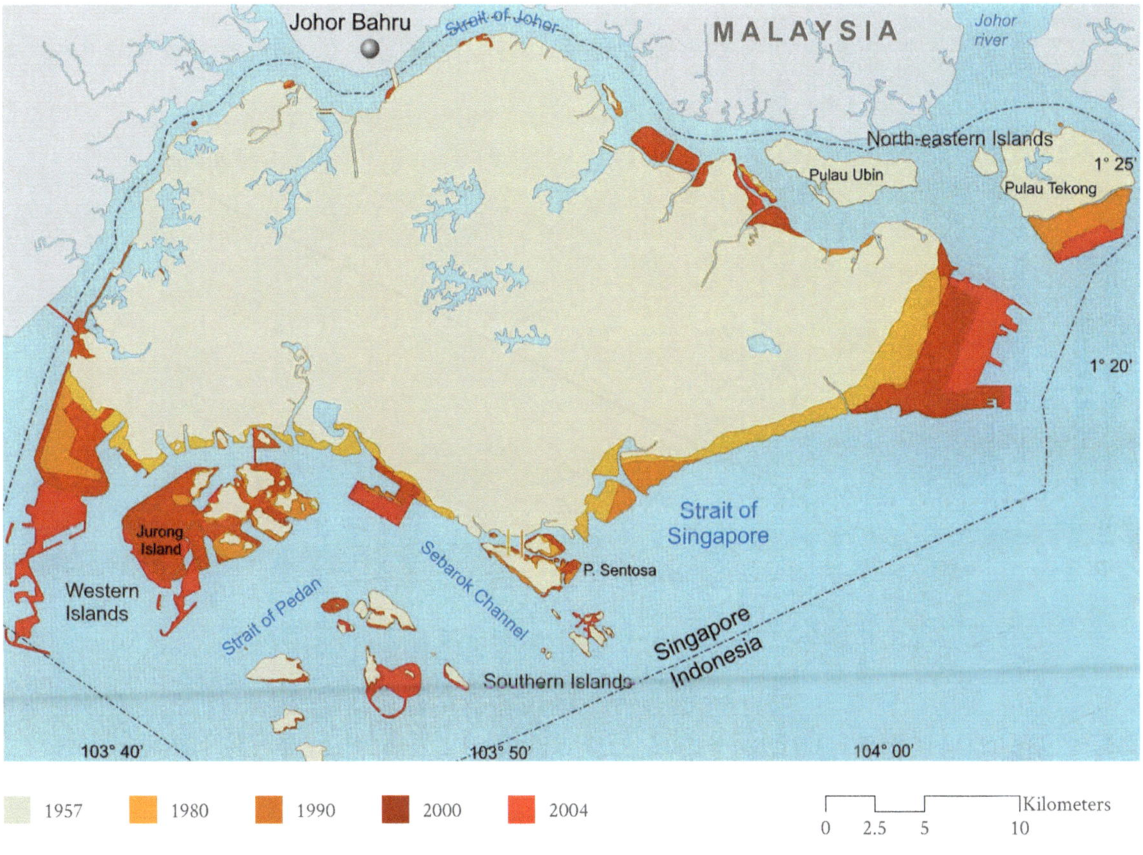

90. LAND RECLAMATION OVER TIME IN SINGAPORE

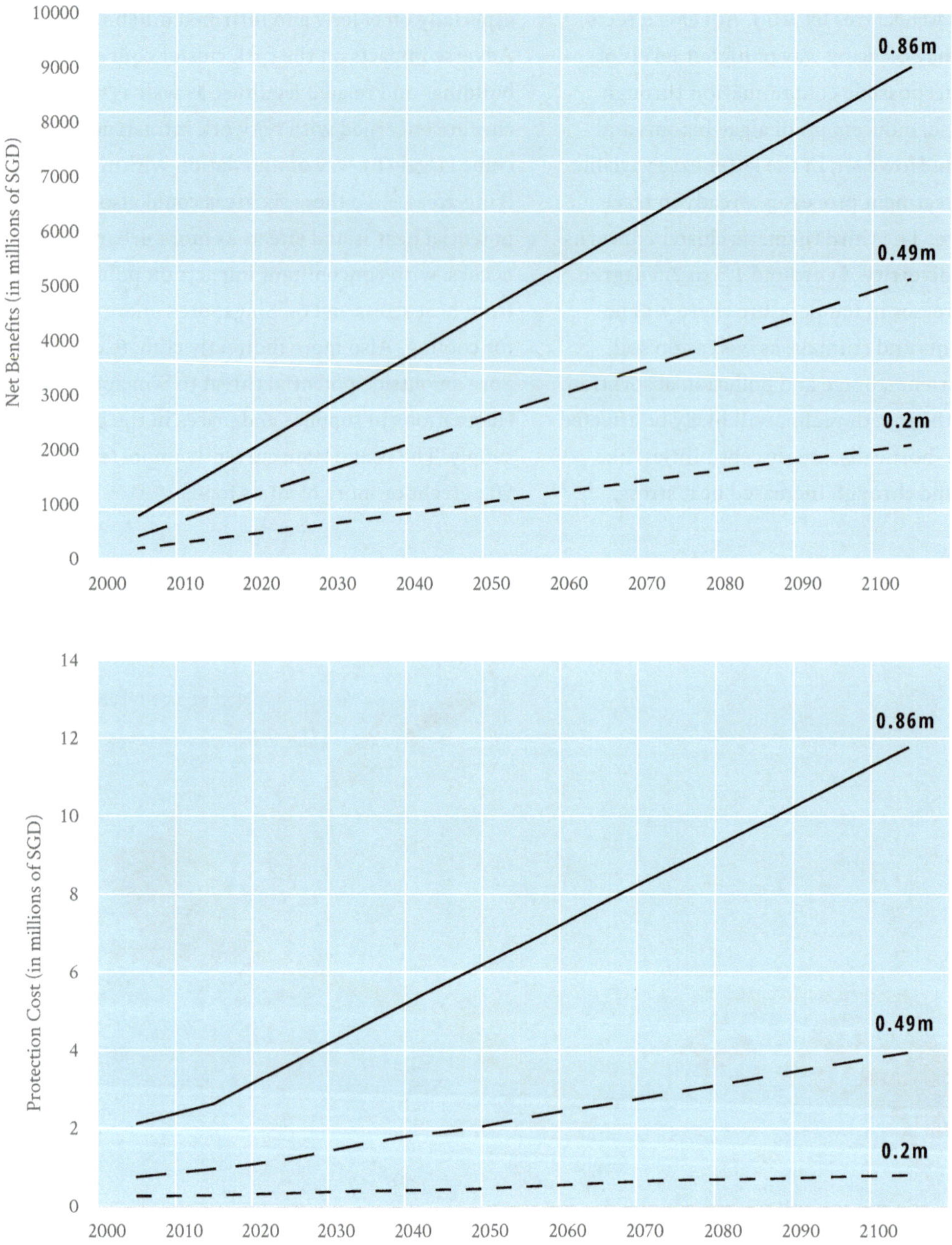

91. ECONOMIC COSTS AND BENEFITS OF SEA-LEVEL RISE PROTECTION

In dealing with the adverse effects of climate change, the IPCC defines 'adaption' as the adjustment in ecological, social, and economic systems in response to actual or expected climate stimuli and their effects. 'Adaptive capacity', in turn, is the ability of such a system or systems to adjust to climate change involving potential damages. In Singapore at least two approaches could be possible. One would be a 'do nothing' approach, whereby land below sea-level rise would be inundated. Overall this would amount to a potential loss of from four to 17 square kilometers of dryland or from 0.6 to 2.7 percent of the island's dryland area.[23] However, this would most likely not be viable for several reasons. First, apart from the essential infrastructure described earlier, one of the vulnerable areas are islands one of which hosts the world's third largest refining centre and a considerable source of Singapore's GDP. Second, other low-lying coastal areas also happen to be of high real-estate value and where the cost of building a seawall protection is outweighed by the loss of property value through inundation. In fact, study shows that for Singapore seawall or similar protection is viable and the best solution because the net present value remains positive. Over time, construction of seawall protection would need to be made upward incrementally, along with maintenance of the seawalls. In this regard, study results show that disruption every decade will likely be severe, though 20 to 30 year intervals would seem to be more feasible with cost variations of 194 and 367 percent respectively.[24] Payment for such operations could be levied across all Singaporeans in some form of a tax and under the rubric that all would benefit. Alternatively, costs could be borne by property owners along affected coastal zones.

Another trans-border phenomenon occurs periodically but with severe impacts. It is the harmful hazing of Singapore from slash and burn farming practices and forest fires for palm olive and other agricultural production, primarily from neighboring Indonesia during the winter months of the year, In fact recently in late November of 2013, the 'Pollution Standard Index (PSI) a usual measure of haze severity reached a record level of 401 units.[25] Now these occurrences are closely monitored and posted on line daily by Singapore's National Environment Agency, where for a 24-hour cycle levels of 0-55 are normal, 56-150 are elevated, 151-250 are high, and levels above 251 are very high and extremely hazardous for public health, often causing schools and other community functions to be closed down. Although it remains to be seen whether haze in Singapore can be prevented permanently, the Indonesian Government has undertaken serious efforts to curb the problems of haze and forest fires. Since 2015, three consecutive years have not seen any substantial hazing. However, strong business ties between the two countries further complicate the situation, and it is in the interest of the two countries to work together amicably to solve the problem.[26]

92. HAZE OVER SINGAPORE

The third potential trans-boundary threat, though not so direct nor regular, could come from risks associated with civilian nuclear power generation close to Singapore. Again this can be seen to stem primarily from Indonesia's nuclear program, as Malaysia though making nods in the nuclear direction have adequate fossil fuel for power generation. Although it is one of the world's largest natural gas producers Indonesia imports its oil and other energy fuel stocks. Consequently it has ramped up its nuclear power plant production with facilities on Madura Island in Eastern Java and on the Miura Peninsula on the northeast coat of Java. Targeting a 26 percent reduction in carbon dioxide emissions by 2020, Indonesia has been deliberately switching away from fossil fuels. Beginning in 1989 with the study that led to the Miura nuclear plant, four other plants are scheduled to open by 2025 aimed at a total of two percent of the nation's electrical production by 2017. Indeed, Indonesia is not alone in Southeast Asia or among ASEAN members in at least exploring if not committing to nuclear plant options. By contrast, in 2007, Singapore was alone in the ASEAN region in expressing caution and concern over safety and security of nuclear power plant facilities.[27] By the same token, its local petrochemical refining capacity is very high, much of it for export. In sum, in a manner similar to 'virtual water' in earlier discussion, these trans-boundary phenomena and effects muddle the almost pristine ecological capsule Singapore has created for itself by being the cleanest and greenest city in all of Asia, if not beyond, with exacting environmental laws, a place where smokers are shunned and where the cost of owning and driving a car are prohibitive.

d. DEALING WITH FUTURE UNCERTAINTY

At the risk of being overly repetitive, with regard to water security, to maintenance of a 'clean and green' island state, and to the idea of a 'City in Nature', Singapore is very well positioned and seems more than likely to succeed. In short, those aspects in this book which are essentially internal and local to Singapore as a circumscribed entity and over which it has more or less full control, are demonstrably well addressed and tractable. More particularly, this is probably strongest in the cases when the issues at hand are or can be framed as being technocratic and resolved pragmatically. However, as pointed out at several junctures in the book, some hyper-objective and virtual aspects of Singapore and, therefore, existing outside its circumscribed boundary, can lead to challenges from the outside to the island state's equilibrium and state of being, or they exist in a manner which probably escapes the dominant technically-inclined modes of thinking and action now in play. This applies to challenges around climate change discussed earlier, as well as to other existential threats to its way of life, such as diminished environmental quality due to the actions from others outside. Also implications of demographic shifts to many more non-Singaporeans in order to maintain high levels of living standards and wealth carry socio-political risks of possible and destructive discord, quite apart from further disjunction between the 'haves' and 'have nots' in the country.

What might or can be done is a matter largely of conjecture, although several courses of action seem to be attractive if not entirely necessary. First, Singapore needs to be seen more clearly, internationally, as resolving its issues. This would include beginning to take concrete steps to 'wall out' and to assimilate sea level rise and to be seen to exercise responsible judgement in reducing accumulation of virtual water resources. This can be done through continued innovation of water treatment processes, like NEWater and desalination, especially in response to increasingly more variable feed water qualities occasioned by climate change. Also Singapore can consider self-imposition of something like a 'virtual water tax' when making trade and other agreements with foreign countries for the importation of food and industrial products. Nevertheless, pursuit of complete domestic water independence, though technically feasible, may not be the right political play, as evidenced by recent rising tensions with Malaysia.[28] Amicable interdependence in trans-border circumstances might will be a superior and more pragmatic strategy. Then too, achievement of greater labor efficiencies, almost across the board, could result in a more balanced and harmonious community of Singaporeans in the future. In turn, these efforts would allow Singapore to take a higher moral tone in the world in pushing others towards better environmental outcomes and to everyone's benefit. Part of this *moral suasion* should also include technical transfer of Singapore's by now considerable knowledge and expertise in managing the water sector.

A second course of action would comprise more sustained and better efforts towards innovation. This would seem most likely to occur in the realm of assemblage and operation of environmental as well as other related technologies. Certainly the recent step taken by Singapore to strengthen tertiary education and research sponsorship are good and necessary steps. The stronger orientation of educational programs towards work-place skills and industry-led research agendas will come into play. This would also seem to point further in the direction of innovation and, with success, a lessening of the need to import migrant workers and at both ends of the socio-economic spectrum.[29] Care in these regards must also be exerted to curtail any claims by others of xenophobic behavior on the part of Singaporeans. Also linked closely to these kinds of issues is maintaining a well-informed and accepting citizenry or having everyone on the proverbial same page, so to speak. This has been one of the successes of many of the programs described here, but will become more difficult, potentially, as socio=political pressures mount and higher degrees of uncertainty about viable futures begin to surface and finally become rife. Here Singapore is certainly a victim of its small size. However, as shown at other times this relatively small size can be turned into a positive asset.

endnotes

1. Yeo, H. 2017. "Building a City in Nature". *CLC Insights*, 24, p.3f.

2. Tan, Kenneth Paul, "The ideology of pragmatism in Singapore: neoliberal globalization and political authoritarianism". *Journal of Contemporary Asia*, 42, no. 1 (2012): 67-92.

3. Singapore General Election Results of 2015 had the People's Action Party (PAB) garnering 69.6 percent of the vote for 82 seats. A gain of 9.72 percent more on the previous General Election.

4. Interview with Mr. Tan Nguan Sen, Chief Sustainability Officer, PUB on 7 February 2017.

5. Leo, David. 1995. *Kiasu, Kiasu: You think What?* (Singapore: Times Editions, 160 pages).

6. A dictionary of Singlish and Singaporean English (https://www.singlishdictionary.com/singlish_k_htm).

7. Savage, V. R. and L. Kong. 1993. "Urban Constraints, Political Imperatives: Environmental 'Design' in Singapore." *Landscape and Urban Planning*, vol 25, pp.37-52.

8. Leitman, J. 2000. *Integrating the Environment in Urban Development: Singapore as a Model of Good Practice.* (Working Paper Series no.7. Washington, D.C. The World Bank), pp.10-11 and World Bank, 2006. *Dealing with Water Scarcity in Singapore: Institution, Strategies and Enforcement.* World Bank Analytical and Advising Assistance Program, China: Addressing Water Scarcity, Background Paper no.4, The World Bank, July.

9. Leitmann, 2000, P.10.

10. Tan, L. H. H. and H. Neo, 2009. "Community in Bloom: Local participation of community gardens in urban Singapore." *Local Environment*, 14:6, pp.536-537.

11. Tortajarda, Celia and Yugal Joshi. 2013."Water Demand Management in Singapore Involving the Public" *Water Resource Management*, vol 27, pp.2729.

12. Soh, E. Y. and Yuen, B. 2005. "Government-aided participation in planning Singapore. *Cities*, vol 23,no 1,pp.34-35.

13. Tan, L. H. 2012. "Revitalizing Singapore's Urban Waterscapes: Active, Beautiful, Clean Waters Programme" *Urban Solutions*, Issue 1, July, PP.14-15.

14. Ho, K. L. 2000. *The Politics of Policy Making in Singapore* (New York: Oxford University Press).

15. IPCC – Intergovernmental Panel on Climate Change (Website).

Palmer, Mathew, Jason Lowe, Dan Bernie. et al.2015.
"Chapter 10: Long-term Projections of Sea Level,
Temperature and Rainfall Change" in Meteorological
Service Singapore, Centre for Climate Research and the
16. UK Met Office, Exeter UK. *Singapore 2nd National Climate
Change Study*, p.2f.

17.
Palmer, p.5.

18.
Palmer, p.18.

All based on Palmer, p.18 and on Marzin, Charlene, Adrian
Hines, James Murphy, et al. 2014. "Executive Summary",
in Meteorological Singapore, Centre for Climate Research
19. and the UK Met Office, Exeter, UK. *Singapore 2nd National
Climate Change Study*, PP.8-9.

20.
Marzin, 2014, p.2.

Bhullar, Lorrean. 2013. "Climate Change Adaption
21. and Water Policy: Lessons from Singapore". *Sustainable
Development*, 21:3, p.158.

22.
All based on Marzin, pp.1-10.

Ng, Wei-Shiwen and Robert Mendelsohn. 2005. "The
23. Impact of Sea Level Rise on Singapore". *Environment and
Development Economics.* April, 10:2, p.213.

24.
Ng and Mendelsohn, pp.211-212.

25. BBC NewsAsia. "Singapore haze hits record high from Indonesian fires". 21 June, 2013.

26. David, Randy. 2013 "Haze over Singapore". *Inquirer.Net*, quoting Irene Ng in paragraph 7, posted online in the Real Singapore (6/19/13).

This vested interest of Singapore in the economic growth involved was described as such by a parliamentarian of the Singapore government.

27. Gunn, Geoffrey. 2008. "Southeast Asia's Looming Nuclear Power Industry". *The Asia-Pacific Journal*, Feb. 1, vol 6, no 2.

28. Tommy Sheng Hao Chai. 2018. 'Water Tensions Between Singapore and Malaysia Begin to Boil Over,' *The News Lens*, July 23 article. https://international.thenewslens.com/article/100407 (accessed 10 January 2019).

29. For instance, the next phase of development of the Singapore University of Technology and Design proposes to focus on skill sets and abilities needed by industry in Singapore.

LIST OF INTERVIEWS

The dates of interviews by person and designation, as well as site visits were as follows.

6 Feb. 2017	Mr. Harry Seah, Director of Technology, PUB.
7 Feb. 2017	Mr. Tan Nguan Sen, Chief Sustainability Officer, PUB.
8 Feb, 2017	Mr. Khew Sin Khoon, CEO, CPG Consultants.
9 Feb. 2017	Mr. Tan Gee Paw, Chairman, PUB.
9 Feb. 2017	Mr. Kenneth Er, CEO, NParks and Dr. Leong Chee Chiew, DCEO, NParks.
9 Feb. 2017	Mr. Richard Hassel and Mr. Wong Mun Summ, Co-Founders, WOHA.
10 Feb. 2017	Mr. Loh Ah Tuan, Former DCEO, NEA.
10 Feb. 2017	Mr. Tobias Baur, Managing Director, Ramboll Studio Dreiseitl Singapore.
10 Feb. 2017	Bukit Timah Reserve site visit.
10 Feb. 2017	Eco-Link@BKE (Bukit Timah Expressway) site visit.
11 Feb. 2017	Alexandra Canal site visit.
11 Feb. 2017	Central Catchment Nature Reserve site visit.
11 Feb. 2017	Singapore Botanic Gardens (Herbarium and Laboratories) site visit.
15 Feb. 2017	Dr. Tan Wee Kiat, CEO, Gardens by the Bay.
22 Jun. 2017	Mr. Yong Wei Hin, Director, Deep Tunnel Sewage System, PUB.
22 Jun. 2017	Mr. Lim Liang Jim, Director, Industrt and Centre for Urban Greenery and Ecology, NParks.
22 Jun. 2017	Dr. Lena Chan, Director, National Biodiversity Centre, Nparks.

22 Jun. 2017	Mr Harry Seah, Chief Technology Officer, PUB
22 Jun. 2017	Sinspring Desalination Plant, Tuas, site visit.
22 Jun. 2017	Bedok NEWater Plant, site visit.
23 Jun. 2017	Mr. Khoo Teng Chye, Executive Director, CLC.
15 Aug. 2017	Ms. Olivia Lum, Founder, Hyflux.
15 Aug. 2017	Mr. Chionh Chye Khye, CLC Fellow; Former DCEO, HDB and Former CEO, BCA.
16 Aug. 2017	Ms. Fun Siew Leng, Assistant Chief Planner, URA.
16 Aug. 2017	Ms. Linda De Mello, Deputy Director, 3P Networks, PUB.
16 Aug. 2017	Mr. Michael Koh, CLC Fellow.
16 Aug. 2017	Gardens by the Bay site visit with Dr. Tan Wee Kiat, CEO, Gardens by the Bay.
17 Aug. 2017	Dr. Cheong Koon Hean, CEO, HDB; Former CEO, URA.
17 Aug. 2017	Mr. Wong Kai Yeng, Former Group Director, URA.
17 Aug. 2017	Prof. Leo Tan, Veteran Marine Biologist; Former Director, Singapore Science Centre.
17 Aug. 2017	Dr. Darren Yeo, Fmr. Chair of National Parks Board; Assist. Prof. Biological Sci. Dept., NUS.
17 Aug. 2017	Lee Kong Chian Natural History Museum, site visit.
18 Aug. 2017	Dr. Liu Thai Ker, Former CEO, HDB and URA; Chairman, CLC.
18 Aug. 2017	Mr Yap Kheng Guan, Former Director, PUB.

IMAGE CREDITS

1. Harvard Graduate School of Design (2018). Satellite Image of Singapore. [image] Available at Google Earth [Accessed 27 Jun. 2018].

2. Urban Redevelopment Authority (1991). Living the next lap: Towards a tropical city of excellence, Singapore.

3A. The Population of Singapore, 1826-2017. (2018). [Computer-aided Diagram] Massachusetts: Harvard Graduate School of Design. Based on Jan Lahmeyer. *Historical Demography-Asia*, Singapore. www.populstat.info/populhome.html

3B. Singapore's Total and Non-resident Population. (2018). [Computer-aided Diagram] Massachusetts: Harvard Graduate School of Design. Based on Singapore Census of Population, Natural Populatation and Total.

4. Economic Intensity of Singapore among Selected Cities. (2018). [Computer-aided Diagram] Massachusetts: Harvard Graduate School of Design. Based on World Bank, 'databank.worldbank.org/data/GDP'; Wikipedia, 'List of Cities by GDP'; OECD, 'Regions GDP(PPP), 2016'; Jegede, 2018, Top 12 Richest Countries in The World. *Trendrr*, December 1.

5. Painting from Iain Mauley. 2010. *Tales of Old Singapore*. Singapore: Earnshaw Books, p.11.

6. Winstedt, R. (1982). Plan of Ancient Singapore. [Sketch] Malaysia: Journal of the Malaysian Branch of the Royal Asiatic Society.

7. Courtesy of the National Museum of Singapore, National Heritage Board

8. Galstaun, A. (1910). Malay Village, Singapore. [Photograph] Singapore: Arshak C Galstaun Collection, National Archives of Singapore.

9. Koninck, Rodolphe De., Drolet, Julie, and Girard, Marc. Singapore : An Atlas of Perpetual Territorial Transformation. Singapore: NUS Press, 2008.

10. Jackson, P. (1822). Plan of the Town of Singapore. [Sketch of Map] Singapore: Government of SIngapore.

11. Maya Jagapal.1991. *Old Singapore: Images of Asia*. Singapore: Oxford University Press, p12.

12. Glass Positive of a Pepper Plantation in Singapore. (1900). [Glass Positive] Singapore: National Museum of Singapore.

13. Guo, W. (2018). Map of Sea Routes to Singapore. [Computer-aided Drawing] Massachusetts: Harvard Graduate School of Design. Based on Theracie. "The Impact of External Events on Singapore". https://www.slideshare.net/theracie/chapter-four

14. Map of the Island of Singapore and its dependencies. (1911). [Map] London: War Office.

15. Rowe, P. (n.d.). Trolley Car on Geyland Road. [Photograph].

16. Corlett RT, (1991). Vegetation. *The Biophysical Environment of Singapore*. Singapore University Press, Singapore.

17. Singapore coffee plantation, late nineteenth century. (1800). [Photograph] Washington: Library of Congress.

18. The Mainichi Newspapers/AFLO.

19. Squatters in Seah Liang Seah Estate at Serangoon Road, Singapore. (1963). [Photograph] Singapore: National Archives of Singapore.

20. After, Frances Loeb Library, Graduate School of Design, Harvard University, Visual Sources.

21. Rowe, P. (n.d.). Night Carts in Operation. [Photograph].

22. Rowe, P. (n.d.). A Turnkey Factory in the Jurong Industrial Park. [Photograph].

23. Urban Redevelopment Authority Singapore.

24. Prime Minister Lee Kuan Yew Planting a Sapling During His Tour of Ulu Pandan Constituency. (1963). [Photograph] Singapore: Ministry of Information and the Arts.

25. Koninck, Rodolphe De. Singapore: Singapore's Permanent Territorial Revolution: Fifty Years in Fifty Maps. Singapore: NUS Press, 2017.

26. National Parks Board.

27A. James Tan.

27B. National Parks Board

28. Howard, E. (1902). The Garden City Concept. [Sketch] United Kingdom: Garden Cities of Tomorrow.

29A. Miller, M. (2018). Letchworth. [Photograph] United Kingdom: Letchworth: The First Garden City, Volume 2.

29B. MacFadyen, D. (1970). Welwyn, United Kingdom. [Photograph] Manchester: Sir Ebenezer Howard and the Town Planning Movement.

30. Urban Redevelopment Authority Singapore.

31. Schinkel, S., Selter, S., & Memhard, M. (n.d.). Berlin: Maps, Plans, Diagrams, Berlin, Germany.

32. Möhring, B., Eberstadt, R. and Peterson, R. (1890). Diagram of the Berlin Metropolis. [Sketch] Munich and London: Shaping the Great City: Modern Architecture in Central Europe, 1890-1937.

33. WOHA.

34. Bennett, W. (1833). American Pastoralism: Richmond from the Hill Above the Waterworks. [Hand-colored Aquatint] Washington: National Gallery of Art.

35. Circuit boards, Atlanta 2004, from the series Intolerable Beauty: Portraits of American Mass Consumption (2003 - 2005), by Chris Jordan.

36. Inness, G. (1856). The Lackawanna Valley. [Oil on Canvas] Washington: National Gallery of Art.

37. Hénard, E. (2018). Bois du Boulogne. [Sketch] Paris.

38. Haussmann, B. (1853). Example of a boulevard created by Haussmann. [Sketch] Paris.

39. Lorenzetti, A. (1338). The Allegory of Good and Bad Government. [Fresco] Siena: Palazzo Pubblico.

40. Rowe, P. (n.d.). Garbatella, Rome.[Photograph].

41. colored image of huts: Galstaun, A. (1900). Farmer's House, Singapore. [Painting] Courtesy of the National University of Singapore Museum Collection.

image of building: A Kampong School Compound. (1948). [Painting] Singapore: Shin Min Public School Collection, National Archives of Singapore.

image of vegetation: Rural Board Facilities. (1952). [Painting] Courtesy of the National Museum of Singapore, National Heritage Board.

image of map: Coleman, G. (1839). Map of the Town and Environs of Singapore from an Actual Survey. [Survey Map] Singapore: National Archives of Singapore.

42. Chris Wainwright. Red Ice 3. 2009.

43. PUB, Singapore's National Water Agency.

44. PUB, Singapore's National Water Agency.

45. PUB, Singapore's National Water Agency.

46. PUB, Singapore's National Water Agency.

47. PUB, Singapore's National Water Agency.

48. PUB, Singapore's National Water Agency.

49. PUB, Singapore's National Water Agency.

50. Innovations in water treatment technology, drawn by Luke Tan.

51. Hyflux Limited

52. PUB, Singapore's National Water Agency.

53. PUB, Singapore's National Water Agency.

54. PUB, Singapore's National Water Agency.

55. Kennedy, A. and Sankey, M. (1897). The Thermal Efficiency of Steam Engines. [Sketch] United Kingdom: Minutes of the Proceedings of the Institution of Civil Engineers.

56. A Stock-Flow Diagram of Singapore's Expenditure on Electricity & Water by Sector and Land Use (2017). Numbers are estimates, not exact figures. Data source: Singapore Energy Statistics 2018; URA Master Plan and PUB, Singapore National Water Agency.

57. PUB, Singapore's National Water Agency.

58. Singapore's Virtual Water Use. (2018). [Computer-aided Diagram] Massachusetts: Harvard Graduate School of Design. Based on Vanham, 2011, p223-225.

59. Botanic Garden. (1989). [Photograph] Singapore: Ministry of Information and the Arts.

60. Plan of the Singapore Botanic Gardens. (2018). [Computer-aided Map] Massachusetts: Harvard Graduate School of Design. Based on map from Data.gov.sg. http://data.gov.sg/.

61. National Parks Board.

62. Plan of Gardens by the Bay. (2018). [Computer-aided Map] Massachusetts: Harvard Graduate School of Design. Based on map from Data.gov.sg. http://data.gov.sg/.

63. Gardens by the Bay.

64. Interior of Cloud Forest. (n.d.). [Photograph] Singapore: Gardens by the Bay.

65. Supertree Grove - Night. (n.d.). [Photograph] Singapore: Gardens by the Bay.

66. Image above by Chriskay, below by LW Yang.

67. Urban Redevelopment Authority Singapore.

68. The Central Catchment and Bukit Timah. (2018). [Computer-aided Map] Massachusetts: Harvard Graduate School of Design. Based on map from Data.gov.sg. http://data.gov.sg/.

69. National Parks Board Singapore.

70. National Parks Board Singapore.

71. Tan, R. (2018). Mangroves at Sungei Buloh Wetland Reserve. [image] Licensed under Attribution-NonCommercial-NoDerivs 2.0 Generic (CC BY-NC-ND 2.0). Available at: https://flickr.com/photos/wildsingapore/13600735844/in/photolist-mHRjVA-bCY9Ak-aEe4X2-6w4HrD-aEhVym-roLUvS-bq4dmf-egDAGC-VYocfr-ayhPcr-aEe538-24P3dsd-XaW6X1-aEhW1y-aEe4Jz-XmQLc7-rD3KoJ-6TF7cR-WPJXbJ-7-uvx8R-aEe6wx-kkdREg-jeAMD-XdHJoe-kkdUUT-etU3XM-pFin83-ThwGzu-XdGAHB-254JPzH-ayhvv2-T7kNK3-afFwU4-4bzMCE-4baE3Y-mHRi2q-aEhVdu-722LDR-kW3Vvn-217THaf-ayhest-etRgVg-egxQzX-jeAUi-qzNdG9-aEhVC5-mHPqDK-q412X6-YWEXXW-8a8fMj [Accessed 22 Jun. 2018].

72. Tan, R. (2018). Labrador shore. [image] Licensed under Attribution-NonCommercial-NoDerivs 2.0 Generic (CC BY-NC-ND 2.0). Available at: https://flickr.com/photos/wildsingapore/421832694/in/photolist-Dh1au-Gn7Drc-23tJ2Nz-GQg6zs-ZRJdCB-ELuCwv-24hahA8-234dZq4-23GhgsS-Fmxrdu-avDwSx-6XEX7g-24fnXsS-249dUHh-EwVTDP-E1H2Ah-RUDzYY-UR2z2K-HaU4ei-21BNgF8-ZNU1kt-25XqELA-avGf5s-247DBaU-G1FwZK-24E9gXe-6XJXwh-25XqAQj-24Wb2oy-23HNMJ9-cfAUZ3-24UiXUo-DNSZtn-Fh4DJv-6XJKQJ-Hgrp58-Gn7FWx-avGg5h-Qjguc4-QCC2mh-24WbfCE-6XJVYJ-5DptZV-fkyD5g-6XES5Z-6XJRGQ-DNT5tc-24Wb37h-KyRaod-DYb7jP/ [Accessed 22 Jun. 2018].

73. Ministry of the Environment and Water Resources.

74. Jimmy Tan.

75. Elmich Pte Ltd.

76. Ling, A. (2017). Bird's-eye view of MacRitchie Reservoir Park with Lim Bo Seng Memorial in the foreground. [image] Available at: https://www.thetallandshortofit.com/drone-flying/macritchiereservoir-park/ [Accessed 17 Jun. 2018].

77. National Parks Board.

78. Data.gov.sg. http://data.gov.sg/.

79. Urban Redevelopment Authority Singapore.
Ministry of National Development Singapore.

80A. Sodhi, Navjot S., Lian Pin Koh, Barry W. Brook, and Peter K. L. Ng. "Southeast Asian Biodiversity: An Impending Disaster." Trends in Ecology & Evolution 19, no. 12 (2004): 654-660. doi:10.1016/j.tree.2004.09.006.

80B. K. L. Ng. "Southeast Asian Biodiversity: An Impending Disaster." Trends in Ecology & Evolution 19, no. 12 (2004): 654-660. doi:10.1016/j.tree.2004.09.006.

81. Barry W. Brook, Navjot S. Sodhi, and Peter K. L. Ng. "Catastrophic Extinctions Follow Deforestation in Singapore." Nature 424, no. 6947 (2003): 420-423.

82. Alongi, Daniel M. "Carbon Sequestration in Mangrove Forests." Carbon Management 3, no. 3 (2012): 313-22.

83. Elmich Pte Ltd (2015). World's Largest Vertical Greenery Project – Singapore ITE Headquarters & College Central. [image] Available at: http://elmich.com/global/worlds-largest-vertical-greenery-project-singapore-ite-headquarters-college-central/ [Accessed 22 Jun. 2018].

84. Bingham-Hall, P. (2016). PARKROYAL on Pickering, Singapore. [image] Available at: https://archdaily.com/800182/interview-with-woha-the-only-way-to-preserve-nature-is-to-integrate-it-into-our-built-environment/58381268e58ece8350000156-interview-with-woha-the-only-way-to-preserve-nature-is-to-integrate-it-into-our-built-environment-photo [Accessed 27 Jun. 2018].

85. Bingham-Hall, P. (2016). [Illustration] Singapore: Garden City, Mega City: Rethinking Cities For The Age Of Global Warming.

86. Urban Redevelopment Authority. Urban Redevelopment Authority. http://www.ura.gov.sg/.

87. Urban Redevelopment Authority (1991). Living the next lap: Towards a tropical city of excellence, Singapore.

88. Tan, L. (2018). From a 'Garden City' to a 'City in Nature'. [Computer-aided Diagram] Singapore: CLC Insights, Issue No. 24.

89. Singapore 2nd National Climate Change Study – Phase 1, Chapter 10 – Long Term Projections of Sea Level, Temperature and Rainfall Change.

90. Koninck, Rodolphe De., Drolet, Julie, and Girard, Marc. Singapore : An Atlas of Perpetual Territorial Transformation. Singapore: NUS Press, 2008.

91. Ng Wei-Shiuen, and Robert Mendelsohn. The Impact of Sea Level Rise on Singapore. Environment and Development Economics 10, no. 2 (2005): p.207-208.

92. Choe, J. (2013). Ghost town. [image] Licensed under Attribution-NoDerivs 2.0 Generic (CC BY-ND 2.0). Available at: https://flickr.com/photos/crazyegg95/12283979973/in/album-72157640452634084/ [Accessed 22 Jun. 2018].

BIBLIOGRAPHY

Allan, J. A. (1998). Virtual water: A Strategic Resource, Global Solutions to Regional Deficits. *Groundwater,* (4), 545-546.

Allison, G., Blackwill, R., Wyne, A., & Kissinger, H. (2012). *Lee Kuan Yew: The Grand Master's Insights on China, the United States, and the World.* Cambridge: MIT Press. Retrieved from http://www.jstor.org/stable/j.ctt5vjp6m

Alongi, D. M. (2012). Carbon sequestration in mangrove forests. *Carbon Management,* 3(3), 313-322. doi:10.4155/cmt.12.20

Alphand, A. (1867–73). *Les Promenades de Paris, Histoire—Description des Embellissements—Dépenses de Création et d'Entretien des Bois de Boulogne et de Vincennes, Champs-Élysées—Parcs—Squares—Boulevards—Places Plantées, Etude sur l'Art des Jardins et Arboretum. Paris,*. 1 vol text and atlas

Babovic, V., Zhang, J., Tay, S. & Wang, X., Pijcke, G., Manocha, N., Li, X., Meshgi, A., Van Gils, J., Minss, T. and Ong, M. (2017). Modelling reservoir water quality under the effects of climate change -Looking at reservoir quality under different climatological and socio-economic projections. *PUB Innovation in Water 9.*

Barnard, T.P. (2014). *Nature Contained: Environmental Histories of Singapore.* Singapore: National University of Singapore Press.

Barnard, T. P., ed. (2014). Introduction. In T. P. Barnard (Ed.), *Nature Contained: Environmental Histories of Singapore. Singapore*: National University of Singapore Press.

Barnard, T.P. & Emmanuel, M. (2014). Tigers of colonial Singapore. In: T. P. Barnard (Ed.), *Nature Contained: Environmental Histories of Singapore.* Singapore: National University of Singapore Press.

Barnard, T. P., (2016). *Nature's Colony: Empire, Nation and Environment in the Singapore Botanic Gardens,* Singapore: NUS Press.

Barrell, J., (1980). *The Dark Side of Landscape.* NY: Cambridge University Press.

Batchelor, P. (1969). The Origin of the Garden City Concept of Urban Form. *The Journal of the Society of Architectural Historians* 28(3), 184-200.

BBC News Asia. (2013, June 21). Singapore haze hits record high from Indonesian fires. *BBC News Asia.* Retrieved from http://www.bbc.com/news/world-asia-22998592

Bermingham, A., (1986). *Landscape and Ideology: The English Rustic Tradition,* 1740-1860. Berkley: University of California Press.

Bhullar, L. (2013). Climate change adaptation and water policy: Lessons from Singapore. *Sustainable Development,* 21(3), 152-159. doi:10.1002/sd.1546

Blue-Green Cities. (n.d.). [Wikipedia]. Retrieved from https://en.wikipedia.org/wiki/Blue-Green_Cities

Bodik, I., & Kubaska, M. (2013). Energy and sustainability of operation of a wastewater treatment plant. *Environment Protection Engineering 39*(2), 15-24. doi:10.5277/EPE130202

Bramwell, D., Hamann, O., Heywood, V. and Synge, H. (1987). *Botanic gardens and the world conservation strategy: proceedings of an international conference, 26-30 November 1985, held at Las Palmas de Gran Canaria.* London; Orlando: Published for IUCN by Academic Press.

Brockway, L. (2002). *Science and Colonial Expansion: The Role of British Royal Botanic Gardens.* London: Yale University Press

Brook, B. W., Sodhi, N. S., & Peter, K. L. N. (2003). Catastrophic Extinctions Follow Deforestation in Singapore. *Nature,* 424(6947), 420. doi:10.1038/nature01795

Busenkell, M., & Cachola Schmal, P. (2012). *WOHA: Breathing Architecture.* Munich: Prestel.

Campbell, S. (1996). Green Cities, Growing Cities, Just Cities? Urban Planning and the Contradictions of Sustainable Development. *Journal of the American Planning Association (Summer).*

Carnemolla, A. (1989). *Il Giardino Analogo, Considerazioni Sull'Architettura dei Giardini.* Rome: Officina Edizioni.

Castells, M., Goh, L., & Kwok, R. Y. (1991). *The Shek Kip Mei Syndrome: Economic Development and Public Housing in Hong Kong and Singapore.* New York: Sage Publications, Ltd.

Cazzato, V., Fagiolo, M. and Giusti, M.A. (1993). *Teatri di Verzura, La Scena del Giardino dal Barocco al Novecento.* Florence: Edifir.

Chang, Y. (2016). Energy and Environmental Policy. In: Lim, L.Y.C. (ed.). *Singapore's Economic Development- Retrospection and Reflections.* Singapore: World Scientific Publishing Co Pte. Ltd.

Chen, D.C., Maksimovic, C., Voulvoulis, N., (2011). Institutional capacity and policy options for integrated urban water management: a Singapore case study. *Water Policy* 13 pp. 53-68

Chew, E. C. T., & Lee, E. (Eds.). (1991). *A History of Singapore.* New York: Oxford University Press.

Chou, C. (2014). Agriculture and the End of Farming in Singapore. In T. P. Barnard (Ed.), *Nature Contained: Environmental Histories of Singapore.* Singapore: National University of Singapore Press.

Chou, L.M. (1998). The Cleaning of Singapore River and the Kallang River Basin: Approaches, Methods, Investments and Benefits. *Ocean & Coastal Management,* 38(2), 133-145.

Cianci, M. G. (2008). *Metafore, Rappresentazioni e Interpretazioni di Paesseggi.* Florence: Alinea Editrice.

Civil Service. (n.d.). [Wikipedia]. Retrieved from https://en.wikipedia.org/wiki/Civil_service

Civil Service - Public Service Commission. (2018). Retrieved from https://www.psc.gov.sg/home

Coats, A. (1969). *The Quest for Plants: A History of the Horticultural Explorers.* London: Studio Vista.

Corbett, J. (2001). Charles Joseph Minard: Mapping Napoleon's March, 1861. *CSISS Classics,* University of California: Centre for Spatially Integrated Social Science

David, R. (2013). Haze over Singapore. *Inquirer.Net.* Quote by MP Irene Ng, originally posted in *The Real Singapore* (June 19, 2013). Retrieved from http://opinion.inquirer.net/55131/haze-over-singapore

Day, J. (2010). Plants, Prayers, and Power: The Story of the First Mediterranean Gardens. In: O'Brien, D. (Ed.) *Gardening Philosophy for Everyone.* Chichester: Wiley-Blackwell.

Delano, S. F. (2004). *Brook Farm: The Dark Side of Utopia.* Cambridge: Harvard University Press.

Del Re, M.C. (1997). *I Giardini del Sogno.* Florence: Angelo Pontecorboli Editore.

Dubbini, R. (1994). *Geografie dello Sguardo, Visione e Paesaggio in Età Moderna.* Milan: Einaudi Editore.

Dumpelmann, S. (2012). *Layered Landscapes: Parks and Gardens in the Metropolis.* In D. Brantz et al. (eds.). Thick Space: Approaches to Metropolitanism. Berlin: Transcript, p213-238.

Fishman, R. (1982). *Urban Utopias in the Twentieth Century: Ebenezer Howard, Frank Lloyd Wright, Le Corbusier.* Cambridge: The MIT Press.

Foglesong, R. E. (1986). *Planning the Capitalist City: The Colonial Era to the 1920s.* Princeton, N.J.: Princeton University Press.

Francis, M. and Hester, R.T. (1990). *The Meaning of Gardens. Idea, Place and Action.* Cambridge: The MIT Press.

Friess D, Richards D, and Phang V. (2016). Mangrove Forests Store High Densities of Carbon across the Tropical Urban Landscape of Singapore. *Urban Ecosystems* 19(2), 795-810.

Furse-Roberts, J. (2005). *Botanic Garden Creation: The Feasibility and Design of the New British Collections.* Reading: University of Reading.

Gardens by the Bay. (n.d.). [Wikipedia]. Retrieved from https://en.wikipedia.org/wiki/Gardens_by_the_Bay

Gardens by the Bay. (n.d.). [Website]. Retrieved from http://www.gardensbythebay.com.sg/

Gross Domestic Product. (n.d.). [Website]. Singapore: Department of Statistics Singapore. Retrieved from www.singstat.gov.sg

Gunn, G. (2008). Southeast Asia's Looming Nuclear Power Industry. *The Asia-Pacific Journal,* 6(2) Retrieved from http://apjjf.org/-Geoffrey-Gunn/2659/article.html

Gura, P.F. (2007). *American Transcendentalism: A History.* New York: Hill and Wang.

Hack, K., Margolin, J.L., and Delaye, K. (2010). *Singapore from Temasek to the 21st Century: reinventing the Global City.* Singapore: NUS Press

Harrison, R.P. (1992). *Forests: The Shadow of Civilization.* Chicago: The University of Chicago Press

Heywood, F.N. and Vernon, H. The Changing Role of the Botanic Gardens. In: Bramwell, D., Hamann, O., Heywood, V. and Synge, H. (Eds.) (1987). *Botanic gardens and the world conservation strategy: proceedings of an international conference, 26-30 November 1985, held at Las Palmas de Gran Canaria.* London; Orlando: Published for IUCN by Academic Press.

Heywood, F.N. and Wyse, P. (1991). *Tropical Botanic Gardens: Their Role in Conservation and Development.* Oxford: Academic Press.

Heutte, F. (1872). A New Concept: The Commercial Botanical Garden. *American Horticulturalist,* 14-17.

Hill, A. W. (1915). The History and Functions of Botanic Gardens. *Annals of the Missouri Botanical Garden,* 2(1/2), 185-240. Retrieved from http://www.jstor.org.ezp-prod1.hul.harvard.edu/stable/2990033

Ho, K. L. (2000). *The Politics of Policy-making in Singapore.* New York: Oxford University Press. Retrieved from http://catalog.hathitrust.org/Record/004138233

Hoekstra, A.Y, & Chapagain, A. (2007). Water Footprints of Nations: Water Use by People as a Function of their Consumption Pattern. *Water Resources Management; an International Journal - Published for the European Water Resources Association (EWRA),* 21(1), 35-48. doi:10.1007/s11269-006-9039-x

Hoekstra, A.Y. and Chapagain, A.K. (2008). *Globalization of Water – Sharing the Planet's Freshwater Resources.* Oxford: Blackwell Publishing.

Holttum, R. E. (1970). *The Historical Significance of Botanic Gardens in South East Asia.* Taxon, 19(5), 707-714. doi:10.2307/1219283

Howard, E. (1965). *Garden Cities of Tomorrow* (1890). Cambridge, Mass.: MIT Press.

Hoyer, J., Dickhaut, W., Kronawitter, I., & Weber, B. (2011). *Water Sensitive Urban Design.* Hamburg: Jovis.

Hui, J. (1995). Environmental Policy and Green Planning. In: Ooi, G.L., ed. 1995. *Environment and the City: Sharing Singapore's Experience and Future Challenges.* Singapore: Times Academic Press for the Institute of Policy Studies.

Hunt, J.D. (1992). *The Pastoral Landscape.* Hannover, London: University Press of New England.

Hyams, E. and MacQuitty, W. (1969). *Great Botanical Gardens of the World.* New York: The Macmillan Company.

H2PC Asia, (n.d.) *Energy Efficient Seawater Desalination in Singapore.* [pdf] Available at http://www.e2singapore.gov.sg/DATA/0/docs/NewsFiles/Energy%20efficient%20desalination.pdf

International visitor arrivals statistics 2014. (2015). Singapore: Singapore Tourism Board. Retrieved from https://www.stb.gov.sg/statistics-and-market-insights/Pages/statistics-Visitor-Arrivals.aspx

IPCC - Intergovernmental Panel on Climate Change [Website]. Retrieved from https://ipcc.ch/

Johnson, D.E. (1985). *Literature on the History of Botany and Botanic Gardens 1730-1840: A Bibliography.* Huntia, Pittsburgh, Pa.: Hunt Institute for Botanical Documentation, Carnegie-Mellon University

Jordan, D. P. (1995). *Transforming Paris: The Life and Labors of Baron Haussmann.* New York: Free Press.

Khew, J. (2016). Moving Towards a City with Nature. *Innovation* 15(1), 33-39

Kiasu (n.d.). [Wikipedia]. Retrieved from https://en.wikipedia.org/wiki/Kiasu

Kiew, R. (2001). *Singapore Botanic Gardens.* Singapore: Landmark.

Laugier, M.A. (1753). *An Essay on Architecture* (trans. W. and A. Herrmann, 1977). Los Angeles: Hennessey & Ingalls

Lee, J. T. T. (2015). We Built This City: Public Participation in Land Use Decisions in Singapore. *Asian Journal of Comparative Law.* 10, (2), 213-234. Research Collection School Of Law.

Lee, K.W. and Zhou, Y. (2009). *Geology of Singapore.* [pdf]. Singapore: Defence Science and Technology Agency. Available at: https://www.researchgate.net/publication/291262201_Geology_of_Singapore_2nd_Edition

Lee, K. Y. (1968). *Opening Speech of the "Keep Singapore Clean" Campaign* (Singapore Conference Hall). Singapore: National Archives of Singapore.

Lee, K.Y. (2011). *Hard Truths to Keep Singapore Going.* Singapore: Straits Times Press

Lee, K. Y. (2012). *From Third World to First: The Singapore Story 1965-2000: Memoirs of Lee Kuan Yew, Vol 2.* Singapore: Marshall Cavendish.

Leitmann, J. (2000). Integrating the Environment in Urban Development: Singapore as a Model of Good Practice. *World Bank Working Paper Series,* (7)

Lenouvel, N., Lafforgue, M., Chevauché, C., & Rhétoré, P. (2014). The Energy Cost of Water Independence: The Case of Singapore. *Water Science and Technology: A Journal of the International Association on Water Pollution Research*, 70(5), 787. doi:10.2166/wst.2014.290

Leo, D. (1995). *Kiasu, Kiasu: You Think What?* Singapore: Times Books International.

Li, X., Zhang, C., Li, W., Ricard, R., Meng, Q., and Zhang, W. (2015). Assessing Street-Level Urban Greenery Using Google Street View and a Modified Green View Index. *Urban Forestry & Urban Greening 14*, 675-685

Lim, H. S., & Lu, X. X. (2016). Sustainable Urban Stormwater Management in the Tropics: An Evaluation of Singapore's ABC Waters Programme. *Journal of Hydrology, 538*, 842-862.

Lin, Y. (2017). *Singapore at the Front Line of Water Innovation.* Straits Times, [online]. March 12. Available at http://www.straitstimes.com/singapore/spore-at-the-front-line-of-water-innovation

Lovelock, C.E., Cahoon D.R., Friess, D.A., Guntenspergen, G.R., Krauss, K.W., Reef, R., Rogers, K., Saunders, M.L., Sidik, F., Swales, A., Saintilan, Thuyen, L.X., and Triet, T. (2015). The Vulnerability of Indo-Pacific Mangrove Forests to Sea-Level Rise. *Nature* 526, 559–563 (22 October 2015). doi:10.1038/nature15538

Machor, J. L. (1987). *Pastoral Cities: Urban Ideals and the Symbolic Landscape of America.* Madison, London.: The University of Wisconsin Press.

Machuca, L., & Fara, V. (2014). Combination of Electrodialysis and Electrodeionization for Treatment of Condensate from Ammonium Nitrate Production. *World Academy of Science, Engineering and Technology International Journal of Energy and Power Engineering*, 8(6), 485-487.

Malone Lee, L.C. and Kushwaha, V., (n.d.) Project "Eco-efficient and Sustainable Urban Infrastructure Development in Asia and Latin America". Case Study "Active, Beautiful and Clean" Waters Programme in Singapore. UNECLAC, UNESCAP.

Mayr, M., Alonso, C., and Rousse, C. (2017). *Blue-Green Network Planning as a Spatial Development and Climate-Resilient Strategy – The Case of Belmopan, Belize.* Caribbean Urban Forum 2017, 15-19 May, Belize City, Belize.

Maresca, P. (1997). *Boschi Sacri e Giardini Incantati.* Florence: Angelo Pontecorboli Editore.

Marx, L. (1964). *The Machine in the Garden : Technology and the Pastoral Ideal in America.* New York: Oxford University Press.

Marzin, C., Hines, A., Murphy, J., Gordon, C., & Jones, R. (2015). *Executive Summary. Singapore 2nd National Climate Change Study.* Meteorological Service Singapore, Centre for Climate Research and UK Met Office

McIntosh, C. (2005). *Gardens of the Gods: Myth, Magic and Meaning.* London, NY: I.B. Tauris.

Miller, M. (2010). *English Garden Cities : An Introduction.* Chicago: University of Chicago Press.

Mitchell, W.J. T. (2002). *Landscape and Power.* Chicago: University of Chicago Press.

Monem, N. K. and Blanche, C. (2007). *Botanic Gardens: A Living History.* London: Black Dog.

Morton, T. (2013). *Hyperobjects: Philosophy and Ecology after the End of the World.* Minneapolis: University of Minnesota Press

Mueller, F. von (1971). *The Objects of a Botanic Garden in Relation to Industries: A Lecture Delivered at the Industrial and Technological Museum.* Melbourne: Mason, Firth & McCutcheon.

Mumford, L. (1946). Garden Cities and the Metropolis: A Reply. *The Journal of Land & Public Utility Economics*, 22(1), 66-69.

Myrie, S. and Arnone, E. (2006). Connecting with Teens: Strategies for Engaging Youth in Botanic Gardens. In: *The Nature of Success: Success for Nature. 6th International Congress on Education in Botanic Gardens*, Oxford, 2006. Available at <http://www.bgci.org/education/1588/>.

National Parks Board (2017). *Parks and Nature Reserves.* Singapore: National Parks Board

Nature Reserves in Singapore. (n.d.). [Wikipedia]. Retrieved from en.wikipedia.prg/wiki/Category:Nature_Reserves_in_Singapore

Neo, B.S., Gwee, J., and Mak, C. (2012). Case Study 1. Growing a city in a garden. In: Gwee, J. ed. 2012. *Case Studies in Public Governance: Building Institutions in Singapore.* Abingdon, Oxon; New York: Routledge

Neo, B.S. (2009). *A City in a Garden - Developing Gardens.* Singapore: Centre for Governance and Leadership, Civil Service College.

Ng, P. K. L., and Tan, L. W. H. (1994). *Rhythm of the Sea: The Life and Times of Labrador Beach.* Singapore: Division of Biology, National University of Singapore.

Ng, W., and Mendelsohn, R. (2005). The Impact of Sea Level Rise on Singapore. *Environment and Development Economics* 10(2), 201-215. doi:10.1017/S1355770X04001706

O'Dempsey, T. (2014). Singapore's Changing Landscape since c.1800. In T. P. Barnard (Ed.), *Nature Contained: Environmental Histories of Singapore.* Singapore: National University of Singapore Press.

Oon, C. 2009. Key step to water adequacy. *Straits Times*, June 24 [online]. Available at https://web.archive.org/web/20090627092606/http://www.straitstimes.com/Breaking+News/Singapore/Story/STIStory_394640.html

Palmer, M., Lowe, J., Bernie, D., Calvert, D., Gohar, L., and Gregory, J. (2015). *Chapter 10: Long Term Projections of Sea Level, Temperature and Rainfall Change. Singapore 2nd National Climate Change Study.* Meteorological Service Singapore, Centre for Climate Research and UK Met Office

Parker, P. M. (2013). *The 2013 Import and Export Market for Printed Maps, Hydrographic Charts, Wall Maps, Topographical Plans, and Globes Excluding Book Form in Singapore.* Singapore: The Icon Group International.

Perry, M., Kong, L., & Yeoh, S. A. B. (1997). *Singapore: A Developmental City State.* New York: John Wiley and Sons. Retrieved from http://scholarbank.nus.edu.sg/handle/10635/79698

Peter, J. (2001). Interview with Mies van der Rohe. In P. Lambert (Ed.), *Mies in America.* New York: Henry N. Abrams.

Pitas, V., Fazekas, B., Banyai, Z., and Karpati, A. (2010). Energy Efficiency of the Municipal Wastewater Treatment. *J Biotechnol.* 150, 163-164.

Plans for the Next Century. (1987, April 27). *Straits Times*

Plappally, A.K., Lienhard, V.J.H. (2012). Energy Requirements for Water Production, Treatment, End Use, Reclamation, and Disposal. *Renewable and Sustainable Energy Reviews* 16(7): 4818-4848.

Population Statistics. (2004). 1999/2004 Populstat Site by Lahmeyer, J. [online]. Available at: http://www.populstat.info/

Population Trends, 2017. (2017). Department of Statistics ingapore, Ministry of Trade & Industry. Retrieved from http://www.singstat.gov.sg/publications/publications-and-papers/population-and-population-structure/population-trends

Potteiger, M. and Purinton J. (1998). *Landscape Narratives: Design Practices for Telling Stories.* New York: John Wiley and Sons.

Press Conference by Prime Minister Lee Kuan Yew at Broadcasting House, Singapore

(August 9, 1965). Broadcasting House, Singapore

Prest, J. (1982). *The Garden of Eden: The Botanic Garden and the Re-Creation of Paradise.* London: Yale University Press.

PUB to Expand Groundwater Monitoring Network. (2016, May 26). *Channel News Asia.* Retrieved from https://www.channelnewsasia.com/news/singapore/pub-to-expand-groundwater-monitoring-network-8038514

Public Utilities Board (2013). *PUB Annual Report 2012/2013. Commemorating Fifty Years of Water: From the First Drop.* Singapore: Public Utilities Board.

Public Utilities Board (2014). *ABC Waters Projects.* Singapore: Public Utilities Board.

Public Utilities Board (2016). *Our Water, Our Future.* Singapore: Public Utilities Board. Retrieved from https://www.pub.gov.sg/Documents/PUBOurWaterOurFuture.pdf

Public Utilities Board (2016). NEWater [online]. Available at: http://www.pub.gov.sg/watersupply/fournationaltaps/newater

Public Utilities Board (n.d.). Alexandra Canal [online]. Available at: https://www.pub.gov.sg/abcwaters/explore/alexandracanal

Pugh, S., (1990). *Reading Landscape: Country, City, Capital.* Manchester: Manchester University Press.

Rayton, R. (2000). *Nature's Government: Science, Imperial Britain, and the 'Improvement' of the World.* London: Yale University Press.

Reddy, M. (2013). Bay South Garden. *ICC Plan Review Service* (2013, August 18)

Richards, D.R. and Friess, D.A. (2017). Rates and Drivers of Mangrove Deforestation in Southeast Asia, 2000-2012. *Proceedings of the National Academy of Sciences* 113(2), 344-349

Richie, A. (1998). *Faust's Metropolis: A History of Berlin.* New York: Carrosll and Graf Publishers, Inc.

Rossi, P.O. (1991). *Roma: Giuda all'Architettura Moderna 1909-1991.* Rome: Editori Laterza.

Rowe, P. G. (1991). *Making a Middle Landscape.* Cambridge, Mass.: The MIT Press.

Sankey Diagram. (n.d.). [Wikipedia]. Retrieved from https://en.wikipedia.org/wiki/Sankey_diagram

Sathiamurthy, E., & Voris, H. K. (2006). Maps of Holocene Sea Level Transgression and Submerged Lakes on the Sunda Shelf. *Natural History Journal of Chulalongkorn University, Supplement 2,* 1-43.

Savage, V. R., & Kong, L. (1993). Urban Constraints, Political Imperatives: Environmental 'Design' in Singapore. *Landscape and Urban Planning,* 25(1), 37-52. doi:10.1016/0169-2046(93)90121-S

Schama, S. (1995). *Landscape and Memory.* New York: Alfred A. Knopf Inc.

Schinz, M. (1985). *Visions of Paradise: Themes and Variations on the Garden.* New York: Stewart, Tabori and Chang.

Schmidt, M. (2008). The Sankey Diagram in Energy and Material Flow Management. *Journal of Industrial Ecology*, 12(1), 82-94. doi:10.1111/j.1530-9290.2008.00004.x

Seah, H., Khoo, K., Chua, J., Toh, D., & Chua, S. (2010). Cost Effective Way to Harvest Estuarine Water: Variable Salinity Desalination Concept. *Aqua - Journal of Water Supply: Research and Technology*, 59(1), 452-458. doi:10.2166/aqua.2010.011

Sheng, Tommy Hao Chai. 2018. 'Water Tensions Between Singapore and Malaysia Begin to Boil Over,' *The News Lens*, July 23 article. https://international.thenewslens.com/article/100407 (accessed 10 January 2019).

Shepherd, J.C. and Jellicoe, G.A. (1986). *Italian Gardens of the Renaissance*. New York: Princeton Architectural Press.

Siemens: Our history in Singapore. (2017). Siemens Historical Institute. Retrieved from https://www.siemens.com/content/dam/webassetpool/mam/tag-siemens-com/smdb/corporate-core/communication-and-gov-affairs/tl/HI/laenderprofile/conversion-pdf-en/028-country-profile-singapore-e-201706.pdf

Siemens to Develop Innovative Seawater Desalination Technology. (2008, June 23). *Siemens Press Release*. Retrieved from http://sg.siemens.com/press/industry/Pages/Siemenstodevelopinnovativeseawaterdesalinationtechnology.aspx

Singapore Exports 1964-2017 (n.d.). Trading Economics. Retrieved from https://tradingeconomics.com/singapore/exports

Soderstorm, M. (2001). *Recreating Eden: A Natural History of Botanical Gardens*. London: Vehicule Press.

Sodhi, N. S., Koh, L. P., Brook, B. W., & Ng, P. K. L. (2004). Southeast Asian Biodiversity: An Impending Disaster. *Trends in Ecology & Evolution*, 19(12), 654-660. doi:10.1016/j.tree.2004.09.006

Sodhi, N., Posa, M., Lee, T., Bickford, D., Koh, L., & Brook, B. (2010). The state and conservation of southeast asian biodiversity. *Biodiversity and Conservation*, 19(2), 317-328. doi:10.1007/s10531-009-9607-5

Soh, E. Y., & Yuen, B. (2006). Government-Aided Participation in Planning Singapore. *Cities*, 23(1), 30-43. doi:10.1016/j.cities.2005.07.011

Soundararajan, K., Ho, M.K. and Su, B. (2014). Sankey Diagram Framework for Energy and Exergy Flows. *Applied Energy 136*, 1035-1042.

Spain, D. (2001). *How Women Saved the City*. Minneapolis, MN.: University of Minnesota Press.

Spens, M. (2007). Deep Explorations into Site/Nonsite: The Work of Gustafson Porter. *Architectural Design*, 77(2), 66-75.

Stuart Gager, C. (1937). Botanic Gardens in Science and Education. Science, *New Series, 85*(2208), 393-399.

Sutcliffe, A. (1984). *Metropolis, 1890-1940*. Chicago: University of Chicago Press.

Tafuri, M. (1996). *Architecture and Utopia*. Cambridge: The MIT Press.

Tagliaventi, G. (1995). *Cittá Giardino: Cento Anni di Teori Modelli Esperieze*. Rome: Gangemi Editore.

Tan, I. (2016). *Beyond Singapore's Blue-Green Matrix: An International Comparison of Blue-Green Systems in Melbourne and Singapore*. Unpublished manuscript.

Tan Keng Yam, T. (2016). *The Singapore Story: Environmental Achievements of This Nation-State over the Past Five Decades*. [online]. Available at: www.unep.org/ourplanet/october-2016

Tan, Kenneth Paul. 2012. 'The ideology of Pragmatism in Singapore: Neoliberal Globalization and Political Authoritarianism.' *Journal of Contemporary Asia*, 42, No.1, p67-92.

Tan, L. H. (2012). Revitalizing Singapore's Urban Waterscapes: Active, Beautiful, Clean Waters Programme. *Urban Solutions Issue 1*, Singapore: Centre for Liveable Cities

Tan, L. H., & Neo, H. (2009). " Community in Bloom": Local Participation of Community Gardens in Urban Singapore. *Local Environment*, 14(6), 529-539. doi:10.1080/13549830902904060

Tan, S.N. (n.d.). [Powerpoint Presentation] *Public Engagement in Planning the Rail Corridor*. Urban Redevelopment Authority

Taylor, N. P. (2014). The Environmental Relevance of Singapore. In T. P. Barnard (Ed.), *Nature Contained: Environmental Histories of Singapore*. Singapore: National University of Singapore Press.

Thompson, P.A. (1972). The Role of the Botanic Garden. *Taxon*, 21 (1), 115-119.

Titman, M. ed., (2013). *The New Pastoralism: Landscape into Architecture*. AD May/June 2013 edition.

Tortajada, C., & Joshi, Y. (2013). Water Demand Management in Singapore: Involving the Public. *Water Resources Management; an International Journal - Published for the European Water Resources Association (EWRA)*, 27(8), 2729-2746. doi:10.1007/s11269-013-0312-5

Tortajada, C., Joshi, Y., & Biswas, A. K. (2013). *The Singapore Water Story: Sustainable Development in an Urban City-state*. New York: Routledge, Taylor & Francis Group.

Turnbull, C. M. (1977). *A History of Singapore, 1819-1988*. Singapore: National University of Singapore.

Vanham, D. (2011). How Much Water Do We Really Use? A Case Study of the City State of Singapore. *Water Science and Technology-Water Supply,* 11(2), 219-228. doi:10.2166/ws.2011.043

Velasco, E., Roth, M., Norford, L., & Molina, L. T. (2016). Does Urban Vegetation Enhance Carbon Sequestration? *Landscape and Urban Planning,* 148, 99-107.

Voris, H. K. (2000). Maps of Pleistocene Sea Levels in Southeast Asia: Shorelines, River Systems and Time Durations. *Journal of Biogeography,* 27(5), 1153-1167. doi:10.1046/j.1365-2699.2000.00489.x

Waylen, K. (2006) *Botanic Gardens: Using Biodiversity to Improve Human Well-Being.* Richmond: Botanic Gardens Conservation International.

Weiss, A.S. (1995). *Mirrors of Infinity, The French Formal Garden and 17th Century Metaphysics.* New York: Princeton Architectural Press

Whittle, T. (1988). *The Plant Hunters: Tales of the Botanist-Explorers who Enriched our Gardens.* New York: PAJ Publications.

Wilson, E. O. (1986). *Biophilia.* Cambridge, Mass.: Harvard University Press

Wilson, J. (2006). Education for Sustainable Development: Guidelines for Action in Botanic Gardens. Richmond: Botanic Gardens Conservation International.

Wong, P.P. (1969). The Surface Configuration of Singapore: A Quantitative Description. *The Journal of Tropical Geography 29,* 64-74

World Bank. (2006). Dealing with Water Scarcity in Singapore: Institution, Strategies and Enforcement. *World Bank Analytical and Advising Assistance Program, China: Addressing Water Scarcity,* Background Paper, (4)

Wright, F.L. Broadacre City: A New Community Plan. *Architectural Record* 99, 243-254.

Wyse Jackson, P.S. and Sutherland, L.A. (2000). International Agenda for Botanic Gardens in Conservation.

Xiu, N., Ignatieva, M. & Konijnendijk Van Den Bosch, C., 2016. The challenges of planning and designing urban green networks in Scandinavian and Chinese cities. *Journal of Architecture and Urbanism,* 40(3), pp.163–176.

Yam, T. W. (1995). *Orchids of the Singapore Botanic Gardens.* Singapore: Singapore Botanic Gardens

Yee, A.T.K., Corlett, R.T., Liew, S.C., and Tan, H.T.W. (2011). The Vegetation of Singapore – An Updated Map. *Gardens' Bulletin Singapore* 63(1&2), 205-212

Yeo, H. and Chng, M. (2017), Building a City in Nature. *CLC Insights 24,* Singapore: Centre for Liveable Cities.

Yok, T. P., Yeo, B., Xi, Y. W. and Seong, L. H.. (2009). *Carbon Storage and Sequestration by Urban Trees in Singapore.* Singapore: Centre for Urban Greenery and Ecology, National Parks Board. Available at: http://botanicgardensshop.sg/shop/carbon-storage-sequestration-by-urban-trees-in-singapore.html

Zhao, C., & Sander, H. A. (2015). Quantifying and mapping the supply of and demand for carbon storage and sequestration service from urban trees. PLOS One 10(8). Doi: 10.1371/journal.pone.0136392

Zhuang, J. (2016). *Marina Bay.* Retrieved from http://eresources.nlb.gov.sg/infopedia/articles/SIP_2016-06-21_160714.html

GLOSSARY

Active, Beautiful, Clean Waters Program (ABC Waters Program)

A program that integrates the environment, water bodies, and the community. This is achieved through the transformation of waterways and reservoirs to create clean and beautiful rivers and lakes, and the shaping of these spaces for community recreation and bonding.

British colonialism

The policy and practice of social, economic and political control by the British Empire over other countries. The Straits Settlements formed in 1826 around possessions on the Straits of Malacca, were comprised of the three trade centres of Singapore, Malacca, Penang, and became a crown colony in 1867 followed by the addition of Labuan in 1912. From 1946 to 1963 Singapore was under British control.

City in a Garden (Singapore)

A philosophy towards urban greening. Proposed by the National Parks Board (NParks) in 1996, it reinvented the "Garden City" strategy to find new strategies and spaces to green Singapore. It was described as "when you are outdoors, you feel that you are in a Garden". This philosophy paved new ways to integrate space, such that the whole was larger than the sum of its parts. Thus, this modified philosophy on Singapore's greening shifted the direction from a "Garden City" to a "City in a Garden".

City in Nature (Singapore)

A proposed philosophy in 2017 by the Centre for Liveable Cities and the National Parks Board (NParks) towards urban greening. It advocates a planning approach to integrate nature into the built environment with the aim to bring nature and biodiversity closer to the people.

City of Gardens and Waters (Singapore)

A vision to integrate water bodies and greenery into Singapore's landscape. Coined by Prime Minister Lee Hsien Loong in 2007, the vision aims to incorporate water infrastructure and greenery into a vibrant and biodiverse city landscape. The most prominent strategy to realise this vision is the Active, Beautiful, Clean Waters (ABC Waters) Programme.

Clean and Green Singapore

The principle to preserve and care for the environment by leading an environmentally conscious lifestyle. It formed the basis of the "Keep Singapore Clean and Green Campaign" in 1968, and the "Clean and Green Week" in 1990. The latter was subsequently renamed "Clean and Green Singapore", and was restructured as a year-long campaign.

Community in Bloom Program (Singapore)

Launched in 2005, this is a nationwide movement aims to foster community spirit by encouraging residents to create and care for their community gardens together.

Concept Plan (Singapore)

One of the two key plans for Singapore's urban planning. Reviewed every ten years, the Concept Plan is a macro-level blueprint, illustrating the broad directions and strategies of the government's land allocation and transportation policy over the next 40 to 50 years. Its vision is then translated into detailed guidelines in the Master Plan.

Deep Tunnel Sewerage System (DTSS)

A superhighway for used water in Singapore. The DTSS satisfies Singapore's long-term water needs in the collection, treatment, reclamation and disposal of used water. It comprises a network of link sewers, which link to two major tunnels that cross Singapore towards three water reclamation plants at the northern (Kranji), eastern (Changi), and western (Tuas) ends of Singapore.

Development Guide Plans (Singapore)

A series of detailed short-term to medium-term future land use plans in a comprehensive review of the Master Plan 1985. Under the Urban Redevelopment Authority (URA), a Development

Guide Plan was drawn for each of the 55 planning areas in Singapore. Together, the 55 Development Guide Plans formed the Master Plan 1998.

Federation of Malaya (described in the book as Malay Federation)

An organisation of nine Malay states and the Straits Settlements of Malacca and Penang. Inaugurated in 1948, it sought to replace the Malayan Union. In 1963, the Federation was merged with Singapore, Sarawak and North Borneo (modern-day Sabah) to form the Federation of Malaysia. Following a series of clashes between the People's Action Party (PAP) government in Singapore and the Alliance leaders in Kuala Lumpur, Singapore separated from Malaysia in 1963 to become a sovereign nation.

Garden City (Singapore)

In 1967, then Prime Minister Lee Kuan Yew introduced this movement for transforming Singapore into a Garden City. The idea was to develop a city with lush greenery and a clean environment for a better quality of life for Singaporeans as well as to attract foreign investments into Singapore.

Garden City Action Committee (Singapore)

A committee set up by then Prime Minister Lee Kuan Yew in 1970 to oversee policies for greening the whole of Singapore and to allow better coordination among government agencies to integrate greenery into the built environment.

Jackson's Plan of Singapore

Jackson's Plan of Singapore refers to Naval Lieutenant Philip Jackson of the Bengal Regiment Artillery under Stamford Raffles. Sometimes referred to as 'Raffles Town Plan'.

Japanese occupation

The occupation by military forces of the former Empire of Japan in Singapore during World War II. After the defeat of the former British colony in the Battle of Singapore, the British surrendered Singapore in 1942. This led to the Japanese Occupation in Singapore that lasted until 1945, during which time Singapore was renamed Syanon.

Land Acquisition Act

Introduced in the 1966, the Land Acquisition Act provides the Singapore government with the legal framework to acquire private land compulsorily at market prices. The primary objective of the act was to make land available readily and inexpensively for a very wide variety of purposes namely, for any public purpose, for public benefit and/or public utility, and for any residential, commercial or industrial purposes.

Marina Barrage

A dam built across the mouth of Marina channel to create Marina Reservoir, Singapore's fifteenth reservoir. This contributed to the expansion of Marina catchment to a sixth of Singapore's land area. The multi-purpose barrage was built not only to provide additional source of water supplyit also serves as flood control and a popular rooftop for recreation activities.

Master Plan (Singapore)

One of the two key plans for Singapore's urban planning. The Master Plan is a statutory land use plan that guides Singapore's development over the next 10 to 15 years. Reviewed every five years, this plan translates the vision and long-term strategies of the Concept Plan into detailed zoning and density parameters for land-use and development.

My Waterway at Punggol

A man-made waterway to transform Punggol into a vibrant Waterfront Town, and is a joint project between the National Parks Board (NParks) and the Housing Development Board (HDB). Together with the Punggol Waterway Park, the 4.2km waterway will traverse Punggol Town, Punggol Promenade, and connect the reservoirs in Sungei Serangoon and Sungei Punggol.

National Parks Board (NParks)

The National Parks Board was derived from the Parks and Recreation Department which was set up under the Ministry of National Development in 1975 after the merger of Singapore Botanic Gardens and Parks & Trees Unit. In 1996, the department was renamed as the National Parks Board where Singapore Botanic Gardens and Singapore's nature reserves came under its custody. It later became the government agency responsible for providing and enhancing greenery for Singapore.

Nature Reserve (Singapore)

A green space which is managed for purposes, including the protection and conservation of its biodiversity, and the researching and dissemination of its aesthetic, historical and scientific significance and knowledge. Conserved under the Parks and Trees Act 2006, Singapore's four Nature Reserves are Bukit Timah Nature Reserve, Central Catchment Nature Reserve, Sungei Buloh Wetland Reserve and Labrador Nature Reserve.

Nature Reserves Ordinance (Singapore)

A law which sets aside nature reserves to preserve the indigenous biodiversity in Singapore, as well as objects and places of aesthetic, historical or scientific interest. It also ensures the conditions and controls for the study of Singapore's biodiversity. Enacted in 1951, the Ordinance legally protected Bukit Timah, Kranji, Pandan, Labrador Cliff, and the Municipal Water Catchment area. It was later replaced by the current National Parks Board Act.

Nature Ways

Routes planted with selected trees and shrubs to encourage movement of fauna between two biodiverse green areas.

Park Connector Network

A network of trail interlinking parks, rivers, canals and residential areas created for people to indulge in various recreational activities such as jogging and cycling. The Park Connector Network started with the intention to use underutilised space along canals and open spaces in residential areas to connect with major parks. Today, it is recognised as an approach to optimise existing space and better integration of green elements as part of the overall garden city strategy.

Park Connector Network

A network of trail interlinking parks, rivers, canals and residential areas created for people to indulge in various recreational activities such as jogging and cycling. The Park Connector Network started with the intention to use underutilised space along canals and open spaces in residential areas to connect with major parks. Today, it is recognised as an approach to optimise existing space and better integration of green elements as part of the overall garden city strategy.

Parks and Trees Act

An act to provide for the planting, maintenance and conservation of trees and plants within national parks, nature reserves, tree conservation areas, and other green spaces in Singapore. Introduced in 1975, the Act was revised in 2017 to grant the National Parks Board (NParks) more regulatory and enforcement power to protect and conserve nature in Singapore. This includes allowing NParks officers to enter homes if the greenery on their premises is a danger to public safety.

Public Utilities Board (PUB), later renamed as the PUB, Singapore's National Water Agency

The national water agency which manages the water supply, water catchment and used water in Singapore. Formed in 1963, it initially served to coordinate the supply of electricity, piped gas and water in Singapore. It was later reconstituted in 2001 as Singapore's national water authority and became a statutory board under the Ministry of Environment and Water Resources (MEWR).

Ring City Concept

An urban planning strategy of a circle of high-density development in Singapore, which flanks the central water catchment area, and a southern development belt from Jurong to Changi. Endorsed by the United Nations (UN) Consultative Review Panel, the strategy was incorporated into the 1971 Concept Plan with technical assistance from the UN.

Sengkang Floating Wetland

A man-made wetland in Punggol Reservoir under the Active, Beautiful, Clean (ABC) Waters Programme of PUB, Singapore's National Water Agency.Constructed on the design theme to "Discover Nature", the wetland improves the water quality and provides a natural habitat for birds and fishes. It comprises of a fixed bridge which connects the Anchorvale Community Centre and Sports Complex, and floating boardwalk to mangroves on the eastern flank of the river.

Singapore Green Plan

It is Singapore's first environmental blueprint which was released in 1992 by then Ministry of Environment to ensure that Singapore could strike a balance between environmental protection and economic growth as it developed. The government later launched the Singapore Green Plan 2012 to include environmental targets which mostly have been met. Instead, the Sustainable Singapore Blueprint was formulated which spells out targets for 2030.

Source-Pathway-Receptor Approach

A method to increase flood protection by adopting catchment-wide solutions. Through flexibility and adaptability to Singapore's full drainage system, this holistic approach considers the drains and canals through which stormwater travels (i.e. "Pathway"), areas that generate stormwater runoff (i.e. "Source") and areas of potential flooding (i.e. "Receptor").

Streetscape Greenery Masterplan (Singapore)

A blueprint initiated by the National Parks Board to accentuate, enhance and revitalize the Garden City, through intensification of streetscapes to create unique identities for clusters of roads at strategic locations. The blueprint spells out planning and design guidelines to ensure better coordination between various agencies such as the housing and transport authorities.

Unaccounted for Water (UFW)

The difference between amount of water delivered to the distribution system and the amount of water sold. UFW mainly consists of water lost through leaks and the under registration of meters. Through comprehensive management of water supply networks, Singapore has ensured its UFW remained at 5%, one of the lowest in the region.

Used water

As part of the sustainable water supply strategy, the term 'used water' is adopted in Singapore instead of the commonly known 'wastewater'. In Singapore, used water is collected and channelled to water reclamation plants, and then treated to international standards of the World Health Organisation.

INDEX

AUTHORS NOTE

PETER G. ROWE is the Raymond Garbe Professor of Architecture and Urban Design at Harvard University's Graduate School of Design and a Harvard University Distinguished Service Professor. He is the author of numerous articles on a variety of topics concerned with cultural, environmental and socio-political aspects of architecture and urban development in various parts of the world. He is also the author, co-author or editor of twenty-one books, including most recently: *Design Thinking in the Digital Age (2017); The Mumbai Metropolitan Region and Palava City: A Brief Account and Evaluation (2017); China's Urban Communities: Concepts, Contexts and Well-Being (2016); and Urban Intensities: Contemporary Housing Types and Territories (2014).*

LIMIN HEE is Director of Research at Singapore's Centre for Liveable Cities (CLC), a knowledge centre for liveable and sustainable cities, where she focusses on research strategies, content development and international collaborations. She is also the author of *Constructing Singapore Public Space (2017); 10 Principles for Liveable High-Density Cities: Lessons from Singapore (2013) and Future Asian Spaces (2012).*

ACKNOWLEDGEMENTS

On a book project like this we were very fortunate to have access to many wonderful former and existing civil servants and members of the private sector, all of whom gave generously of their time. At the PUB, Singapore's National Water Agency, in particular this included: Harry Seah, Tan Nguan Sen, Tan Gee Paw, Yong Wei Hin, Linda De Mello and Yap Keng Guan. Likewise at the National Parks Board (Nparks) our special thanks go to Kenneth Er, Leong Chee Chiew, Lena Chan, Lim Liang Jim and staff members at the laboratories, field offices and Herbarium who made the Botanic Gardens and site visits to reserves so engaging. At the Housing and Development Board (HDB) we were graciously received by Cheong Koon Hean who gave us a fascinating account of the interaction among government agencies and especially collaboration with the PUB and the Urban Redevelopment Authority (URA). At Gardens by the Bay there was never a dull moment in the capable and entertaining hands of Tan Wee Kiat and some of his staff. Among our neighbors at the Urban Redevelopment Authority, Fun Siew Leng and Wong Kai Yeng were very informative. The natural history of Singapore came alive for us through the commentary and discussion, especially with Leo Tan and also with Darren Yeo of the Biological Sciences Department of the National University of Singapore. With his passion for butterflies Khew Sin Khoon at CPG Corporation regaled us with accounts about the relationships of Singaporeans to their natural circumstances. The former Director-General of the National Environment Agency (NEA), Loh Ah Tuan was most helpful in having us understand issues related to river clean ups. Closer to home, at the Centre for Livable Cities, Khoo Teng Chye presided over our conduct with grace, good humor and timely interventions, Fellows Chionh Chye Khye and Michael Koh gave generously of their time and insights. Also we would have been lost without the constant guiding help of Jin Yi Kuang and Joanne Khew. Commenting on almost everything in his inimitable and straight-forward way was Liu Thai Ker. In the private sector, Olivia Lum gave us a dazzling account of the formation of Hyflux; Tobias Baur was both forthright and forthcoming from Ramboll Studio Dreiseitl Singapore and our professional friends Richard Hassel and Wong Mun Summ of WOHA spoke eloquently about public-private co-operation in developing Singapore's broader 'green' vision. Finally but not least, creation of books like this involve many willing participants. At Harvard's Graduate School of Design special thanks go to Hanne van den Berg and her partner Tess Stribos, as well as to ChengHe Guan, Francesca Forlini, Zhangkan Zhou, Luke Tan and the wonderful team of Yun Fu and Wenting Guo. Along with this generous assistance and guidance, any lapses and mistakes are of our own making.

Peter G. Rowe and Limin Hee.

Cambridge, Massachusetts and Singapore,
January 2018.

GPSR Compliance

The European Union's (EU) General Product Safety Regulation (GPSR) is a set of rules that requires consumer products to be safe and our obligations to ensure this.

If you have any concerns about our products, you can contact us on ProductSafety@springernature.com

In case Publisher is established outside the EU, the EU authorized representative is:

Springer Nature Customer Service Center GmbH
Europaplatz 3
69115 Heidelberg, Germany

Batch number: 10845100

Printed by Printforce, the Netherlands